Serious Managers Guide to AI-Enabled Organizations

The Essential Guide to AI Operating Models, Workforce Transformation, and Scalable Governance

Jazper Carter

Cybersoft Publishing LLC

Fort Washington, MD 20744

Drclaude.net

Second Edition March 2025

Table of Contents

Introduction

Why This Book Exists and Why You Need It Now

Most organizations never consciously chose to become "AI-enabled enterprises." It happened quietly, almost accidentally. A pilot chatbot here. A vendor tool with a recommendation engine is available. A forecasting model someone built during a hackathon that somehow made its way into a production workflow. Before long, AI wasn't a future ambition; it was already threaded through customer service, HR, finance, operations, and IT. And then, inevitably, a senior leader asked the question that stops every manager in their tracks:

"Are we sure this thing is safe?"

That question is the moment AI stops being an experiment and becomes your responsibility.

This book is written for people who live within the blast radius of that question: managers, directors, product owners, IT leaders, transformation executives, and operational decision-makers who are accountable for outcomes, even when they are not the ones writing the code. You sit at the intersection of ambition and risk between executives who want AI-powered results, technical teams eager to ship, regulators tightening expectations, and users who assume whatever you deploy is trustworthy.

You don't need another abstract ethics manifesto. You don't need a dense engineering textbook. You need something in between: **a practical, manager-ready guide to building AI-enabled organizations that are fast, safe, and sustainable.**

That is what this book delivers.

The Hidden Reality: AI Changes the Nature of Work

AI does not behave like traditional software. It learns, drifts, adapts, and sometimes fails in subtle, silent, and systemic ways. It can amplify small data issues into large operational problems. It can make decisions that look correct until they suddenly aren't. It can create efficiency breakthroughs and reputational disasters with equal ease.

Most importantly, AI changes the shape of work.

It blurs the boundary between human judgment and machine output. It shifts decision-making from deterministic rules to probabilistic predictions. It forces organizations to rethink ownership, accountability, and information flow. And it exposes a truth many leaders are only now beginning to understand:

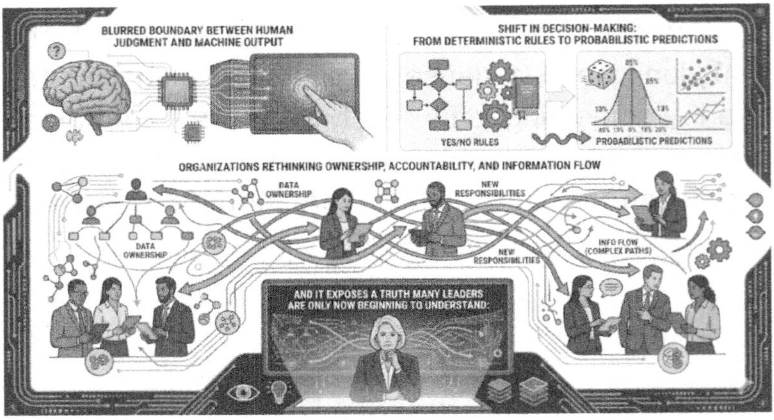

AI is not a technology problem.
It is an operating model problem.

Organizations don't struggle because they lack models. They struggle because they lack clarity about who owns what, how decisions are made, how risks are managed, and how humans and machines should work together.

This book helps you build that clarity.

Why Managers Matter More Than Ever

AI success is no longer determined by who has the best algorithms. It is determined by who has the best managers and leaders who can translate strategy into structure, principles into practice, and innovation into safe, repeatable workflows.

Managers are the ones who:

- Decide which AI use cases move forward
- Define the boundaries between automation and human oversight
- Ensure data is governed, monitored, and refreshed
- Build the cadences that keep models healthy
- Create the conditions for teams to innovate without chaos
- Protect the organization from silent, compounding risk

In short, AI maturity is a management capability, not a technical one.

This book gives you the frameworks, language, and tools to lead confidently, even if you are not a data scientist or engineer.

The New Manager's Challenge: Speed Without Chaos

Every organization adopting AI faces the same tension:

Move fast enough to stay competitive. Move carefully enough to stay safe.

Most fail on one side or the other.

Some move too slowly, paralyzed by uncertainty, committees, and fear of risk. Others move too quickly, deploying models without monitoring, governance, or clear ownership, only to find themselves firefighting issues that could have been prevented with a few hours of planning.

This book shows you how to escape that trap.

You will learn how to design operating models that scale, guardrails that protect without slowing innovation, and human-in-the-loop systems that keep humans in command of high-stakes decisions. You will learn how to structure teams, define roles, build cadences, and create the minimal documentation needed to prevent the most common failure modes.

The goal is not bureaucracy. The goal is **operational clarity,** the kind that reduces friction, accelerates delivery, and builds trust across the organization.

What This Book Will Help You Do

By the time you finish this book, you will know how to:

- Build AI-enabled teams with clear ownership and decision rights
- Design human-in-the-loop workflows that protect safety and ethics
- Choose the right operating model for your organization's maturity
- Prevent model drift, data degradation, and silent failures
- Establish monitoring, escalation, and incident response processes
- Create stage-gates that stop pilots from becoming accidental production systems
- Budget for AI maintenance and avoid hidden cost traps
- Communicate risk, value, and tradeoffs to executives
- Build a culture where responsible AI becomes part of everyday professionalism

You will not become a machine-learning engineer. You will become something far more valuable: **a leader who can make AI work in the real world.**

A Final Thought Before You Begin

AI is not going away. It will not slow down. It will not wait for your organization to feel ready. The question is not whether you will adopt AI, but whether you will adopt it intentionally or accidentally.

This book is your chance to choose intention.

You are about to learn how to build AI-enabled systems that are safe, scalable, and aligned with your mission. You are about to gain the tools to lead with confidence in a landscape where uncertainty is the norm. And you are about to join the small but growing group of managers who understand that AI is not just a technology; it is a transformation in how work gets done.

Turn the page. Your organization's AI future starts with You.

---

1 Reimagining How Work Is Organized for AI

1.1 Scenario: The Pilot That Outgrew Its Team

Eighteen months ago, a regional health system's VP of Engineering approved a modest pilot: a machine-learning model to flag patients at elevated risk of readmission. Two data scientists, one clinical informaticist, and a borrowed product manager stood it up in twelve weeks. The model worked. Leadership praised it at an all-hands. Then the requests came. Could the same model be applied to the orthopedic population? Could it feed the care management dashboard? Could it be adapted for post-surgical follow-up? The original team, still technically a pilot, suddenly owned a production system serving six clinical workflows, two external vendor integrations, and a regulatory reporting requirement. Model drift appeared in week twenty-two. There was no monitoring process. There was no owner. There was no budget line. The VP spent three months firefighting what should have been a thirty-minute governance conversation at the outset. When she finally stood up a proper operating model with clear ownership, a recurring maintenance budget, and stage-gate checkpoints, average time-to-value for new AI initiatives dropped by half within two quarters.

1.2 Why Operating Models Matter Now

Most organizations do not have an AI strategy problem. They have an AI operating model problem. Executives approve pilots. Data science teams build models and products and ship them. And then, quietly, the accumulation of undocumented models, orphaned pipelines, and undefined ownership creates an invisible technical and organizational debt that compounds faster than anyone expects.

An operating model is the set of decisions an organization makes explicitly or by default about how work gets done, who owns what, and how resources flow. In traditional software delivery, operating models evolved over decades. Agile, DevOps, and platform engineering: each represented a renegotiation of how teams are structured, how releases are governed, and how infrastructure is shared. AI introduces a new renegotiation. The differences are not cosmetic.

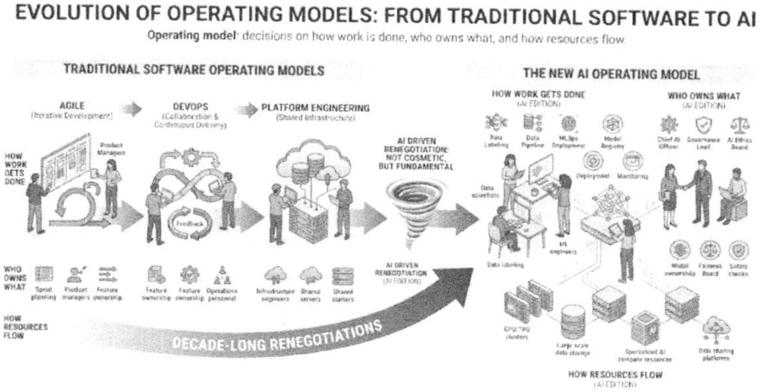

AI systems behave differently from traditional software in three critical ways. First, they degrade over time without any code changes, a phenomenon called model drift, which means ownership cannot end at deployment. Second, their outputs are probabilistic rather than deterministic, which means quality assurance requires different methods than standard QA testing. Third, the datasets that power them are themselves governed assets, subject to privacy regulations and licensing constraints, and to representational risks that traditional software does not carry. None of these realities fit neatly into a standard software delivery operating model.

For the VP of Engineering at a health system, the Director of Product at a federal agency, or the CIO at a mid-market insurer, the question of operating model is not academic. It determines whether AI delivers workflow-level impact or remains an expensive collection of proofs of concept. Getting the operating model right is the difference between AI as a capability and AI as a cost center.

This chapter gives you a practical framework for making operating model decisions that align with your organization's patterns, migrating from ad hoc to a structured approach, and ensuring that governance mechanisms preserve your investment's value over time. The goal is not a perfect org chart. It is operational clarity: every person in your organization who touches AI knows what they own, what they owe, and what success looks like.

1.3 Three Operating Patterns for AI

There is no single correct way to organize AI work in an enterprise. What works for a financial institution with a centralized risk function will not work for a healthcare network with decentralized clinical departments. What works for a federal agency with a unified data infrastructure will not work for a retailer with a dozen regional operations. The goal is not to find the universally correct model; it is to find the model that matches your current scale, risk tolerance, and organizational culture, with a clear line of sight to the model you will need next.

Three operating patterns cover the range of what most enterprises need. Each has a distinct logic, a distinct set of roles, and a distinct set of failure modes. Understanding all three, including where they break down, is what allows a manager to make a mission-aligned choice rather than an imitative one.

Centralized Capability

In the centralized capability pattern, AI work is organized into a single shared function often called a Center of Excellence, AI Platform team, or Data and AI division that serves the rest of the organization. Business units submit requests or participate in joint discovery, but the technical work model development, infrastructure management, data pipeline engineering, and model monitoring lives in one place.

This pattern has real advantages, especially early in an organization's AI journey. It concentrates on scarce talent. It avoids duplication of infrastructure. It creates a natural governance checkpoint, since every model passes through the same team. It also simplifies regulatory compliance, because policies can be enforced at the platform level rather than negotiated with each business unit separately.

The failure mode is throughput. A centralized function becomes a bottleneck as demand scales. Business units that cannot secure capacity from the central team begin building their own shadow functions, sometimes with less rigorous practices, less robust infrastructure, and reduced visibility into the governance layer. The centralized model also tends to create distance between the AI team and the business context, which increases the risk that models are technically sound but operationally irrelevant.

The centralized capability pattern is appropriate when: the organization is early in its AI maturity; the number of active AI initiatives is small enough to be managed by a single team; regulatory risk is high enough to justify centralized oversight; or the organization is investing in foundational infrastructure that will eventually support multiple teams.

Hybrid Product-Platform

In the hybrid product-platform pattern, AI infrastructure compute, data pipelines, model registry, and monitoring tooling are maintained by a central platform team. At the same time, AI product development is distributed across business units or domain-level product teams. The platform provides shared services and sets standards. Product teams build, own, and maintain the models and workflows that serve their domain.

This is the pattern that most mature AI organizations converge on, and for good reason. It preserves the governance and efficiency benefits of centralization at the infrastructure layer, where they matter most, while enabling the speed and domain-knowledge advantages of distributed teams at the product layer. A single-platform team can support dozens of product-level AI initiatives without each initiative having to reinvent data pipelines or monitoring infrastructure.

The failure mode is coordination. When platform and product teams have different priorities, different release cadences, and different success metrics, friction accumulates. Platform teams build capabilities that product teams cannot consume at the pace they need. Product teams build workarounds that undermine the platform's integrity. The solution is not better technology; it is better governance: clear interface contracts between the platform and the product, shared visibility into the roadmap, and an explicit escalation path when priorities conflict.

The hybrid product-platform pattern is appropriate when the organization has multiple active AI product teams or domains, has already made foundational infrastructure investments, has sufficient platform engineering capacity to support the product layer, and has governance mechanisms to manage platform-product interfaces.

Distributed Agentic Networks

The distributed agentic pattern represents the leading edge of AI operating models. In this pattern, AI capabilities are not concentrated in a single function or organized around a product-platform interface. Instead, AI agents, autonomous systems capable of taking actions across tools, data sources, and workflows, are embedded throughout the organization, often communicating with each other and operating within defined boundaries set by a governance layer.

This pattern is beginning to emerge in organizations that have high AI maturity, robust data infrastructure, and a governance framework capable of managing autonomous decision-making. It is not a destination for most organizations reading this book today, but it is the destination most organizations are moving toward over a multi-year horizon. Understanding its structure now allows managers to make architectural decisions about data access, agent design, and governance tooling without foreclosing the distributed agentic path.

The failure mode is accountability. When agents take actions autonomously, tracing a downstream outcome to a responsible human owner becomes difficult without deliberate design. The organizations that navigate this pattern successfully are those that invest heavily in audit logging, human-in-the-loop escalation mechanisms, and clear accountability frameworks before scaling autonomous capabilities. Responsible by design is not a constraint on the distributed agentic model; it is the foundation that makes it viable.

The distributed agentic pattern is appropriate when: the organization has high AI maturity and a functioning governance infrastructure, the use cases are well-defined with clear success criteria and failure boundaries, audit and compliance requirements can be met through technical controls rather than human review at every step, and the organizational trust and cultural alignment needed to operate autonomous systems safely.

Diagram 1.1 – Operating Model Comparison

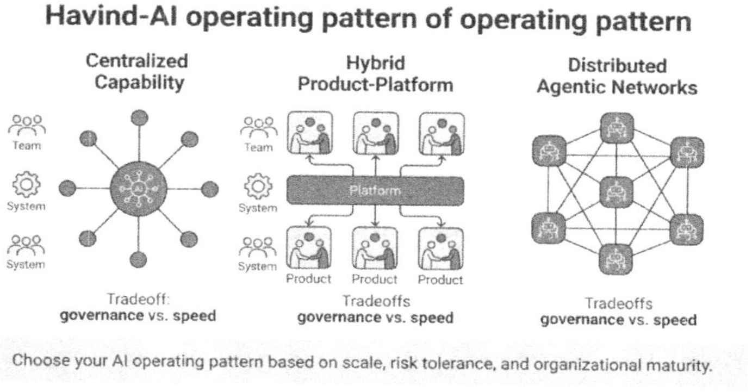

1.4 Choosing the Right Pattern

Choosing an operating pattern is not a values exercise. It is an assessment of your current organizational reality against a set of concrete criteria. Four factors determine which pattern is appropriate at a given moment: scale signals, risk tolerance, cost model, and organizational readiness.

Scale signals are the most direct indicator. If your organization has fewer than five active AI initiatives, a centralized capability model is almost always the right choice, as the efficiency gains from distribution do not justify the coordination costs. If you have five to twenty active initiatives spanning multiple domains, the hybrid product-platform model is typically most effective. If you have more than twenty active initiatives with autonomous decision-making requirements, you are likely ready to begin exploring distributed agentic patterns in bounded domains.

Risk tolerance shapes how you structure governance, regardless of the pattern you choose. High-stakes domains, clinical decision support, credit underwriting, fraud detection, and benefits eligibility require heavier human review and more restrictive approval thresholds, regardless of operating pattern. This is not a reason to centralize; it is a reason to invest in governance tooling that scales with your chosen operating pattern. The pattern governs who builds and owns the work. Governance governs how that work is reviewed, approved, and monitored.

The cost model question is often underweight. AI systems have operating costs that traditional software does not: inference compute, model monitoring infrastructure, data licensing, and the human time required for ongoing model review and retraining. Before choosing a pattern that distributes AI ownership across multiple teams, verify that your budget model can support team-level recurring maintenance costs, not just at the central platform level. Teams that build models without a maintenance budget will inevitably prioritize maintenance, with predictable consequences for model quality and regulatory compliance.

Organizational readiness is the hardest factor to assess honestly. The hybrid product-platform pattern requires product teams that understand AI well enough to make responsible product decisions. The distributed agentic pattern requires governance infrastructure that most organizations do not yet have. Choosing a more advanced pattern than your organization can operate is a common failure mode. A centralized model that works is better than a distributed model that doesn't. The pilot-to-production pathway should match the organizational maturity that exists, not the maturity you wish you had.

Diagram 1.2 – Decision Matrix for Pattern Selection

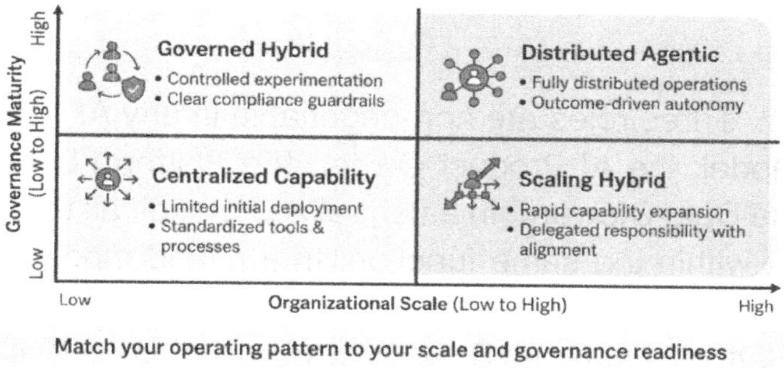

1.5 The Manager's Operating Playbook

Operating patterns are structural choices. The operating playbook is what makes them real. A team can adopt the hybrid product-platform pattern on paper and still operate like an ad hoc pilot culture if it lacks the roles, cadences, and artifacts that give the structure its teeth. This section gives you the minimal viable operating system for an AI-enabled team: the roles you need to define, the meetings you need to run, and the documents you need to maintain.

The word minimal is intentional. Every process has a cost in time, attention, and organizational friction. The goal is not to create a bureaucracy. It is to create the lightest-weight governance infrastructure that prevents the most common failure modes: orphaned models, unclear ownership, unmonitored drift, and invisible costs. Start with the minimum. Add the process only when a specific failure mode justifies the additional cost.

Role Charters

Three roles are non-negotiable in any AI operating model: the AI Product Owner, the Platform Lead, and the Data Steward. In a centralized model, all three may sit within the same function. In a hybrid model, the AI Product Owner may sit within a business unit while the Platform Lead and Data Steward sit in the central team. In a distributed model, each domain may have its own AI Product Owner and Data Steward drawing on a central platform function. The roles themselves do not change; their placement in the org does.

The AI Product Owner is accountable for delivering end-to-end value for an AI system. This is not the same as the traditional product owner role. An AI Product Owner must understand not just what the system should do, but also how to evaluate whether it is doing it, which requires a working understanding of model evaluation metrics, data quality indicators, and the specific failure modes of the AI approach being used. They own the business case, the success metrics, and the decision about when a model is ready for production and when it should be taken offline.

The Platform Lead is accountable for the shared infrastructure on which AI products run, including compute environments, data pipelines, model registries, monitoring tooling, and CI/CD pipelines. In a small organization, this may be one person; in a large organization, it may be a team. The Platform Lead's primary obligation is reliability and standardization: ensuring that AI products can be built, deployed, and monitored using shared tools without each team needing to reinvent the infrastructure layer.

The Data Steward is accountable for the data assets consumed by AI systems. This role sits at the intersection of data governance, privacy regulation, and AI quality. The Data Steward must ensure that training data is appropriately licensed, representative of the population the model will serve, documented in a model card, and compliant with applicable privacy regulations. In regulated industries such as healthcare, financial services, and the federal government, this role is not optional. It is the single point of accountability for the data-related risks that regulators will ask about first.

Role charters should be documented, not just discussed. A one-page charter for each role covering scope of accountability, decision rights, escalation path, and key interfaces is the minimum artifact—verbal agreements about who owns what dissolve under pressure. Written charters hold.

Diagram 1.3 – Role Charter Interaction Map

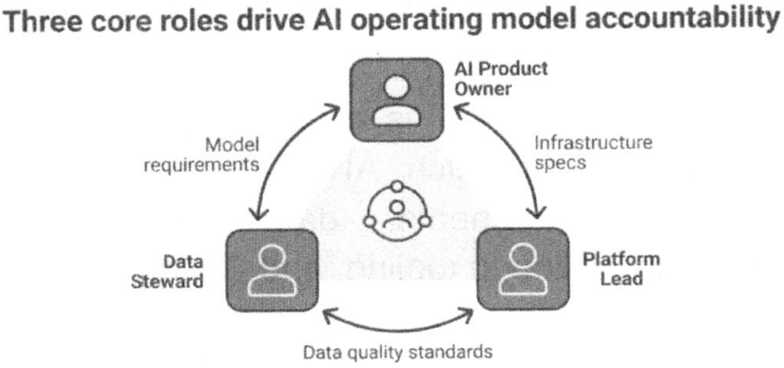

Three core roles drive AI operating model accountability through clear handoffs.

Cadence Design

Operating cadences are the heartbeat of an AI operating model. Without regular, structured touchpoints, ownership diffuses, problems accumulate invisibly, and the governance mechanisms that exist on paper stop functioning in practice. Three cadences cover the essential ground: the weekly model review, the monthly impact review, and the incident postmortem.

The weekly model review is a 15- to 30-minute standing meeting for each active AI system in production. Attendees are the AI Product Owner and whoever owns model monitoring for that system. The agenda is simple: review the monitoring dashboard, flag any drift indicators or anomalies, confirm that inference volumes and latency are within expected ranges, and surface any incoming data quality issues from the Data Steward. This is not a strategy meeting. It is a maintenance check. Its value is in the discipline it creates: no model in production goes more than a week without a human reviewing its health.

The monthly impact review is a broader conversation about whether AI systems deliver the business value they were built to deliver. This meeting includes the AI Product Owner, the relevant business stakeholder, and optionally the Platform Lead. The agenda covers: current performance against target metrics, any workflow-level impact issues that have surfaced since the last review, upcoming model changes or retraining plans, and budget status relative to the recurring maintenance allocation. This is where the decision to invest more, hold steady, or deprecate a model gets made.

The incident postmortem runs whenever an AI system fails in a way that causes a business impact, a model goes offline, produces systematically wrong outputs, or triggers a compliance concern. Postmortems are not blame exercises. They are learning exercises. A well-run postmortem produces a root cause analysis, a set of corrective actions with owners and deadlines, and a process update that reduces the likelihood of the same failure recurring. Over time, a library of postmortem findings is one of the most valuable governance artifacts an AI-enabled organization can maintain.

The Minimal Artifact Set

Documentation is a contested topic in technology organizations. Agile culture rightly pushed back on documentation for its own sake. But AI systems pose risks to users, the organization, and the people whose data powers the model, and require a minimum level of documentation to be managed responsibly. The minimal artifact set is not about compliance theater. It is about ensuring that the people who need to make decisions about an AI system have the information they need to make them correctly.

The Operating Model Canvas is a one-page document that captures, for each AI system: the business problem being solved, the AI approach being used, the data sources and their governance status, the deployment environment, the ownership structure, the success metrics, and the monitoring and escalation plan. It is not a detailed technical specification. It is the minimum information a new team member, an auditor, or a senior leader needs to understand what the system does and who is accountable for it.

The Stage-Gate Form is used at each major milestone in a model's development lifecycle, typically at the transition from prototype to pilot, from pilot to limited production, and from limited production to full deployment. It captures the evidence that the system meets the criteria for the next stage: technical quality metrics, business validation findings, governance approvals, and risk assessments. The stage-gate form is the organizational mechanism that prevents a pilot from becoming a production system before it is ready or before the governance infrastructure needed to support it is in place.

The Model Card Pointer is a lightweight reference document that points to the technical model card, a structured summary of a model's purpose, training data, performance characteristics, and known limitations, maintained by the data science team. The pointer ensures that even stakeholders who will never read a technical model card know it exists, know where to find it, and know who maintains it. In regulated industries, the ability to produce a model card on request is increasingly an expectation, not a courtesy.

1.6 Migration Playbook Pilot to Platform

One of the most consistent patterns in AI program failure is the absence of a defined pilot-to-production pathway. Organizations invest in pilots without a clear mechanism for deciding when and how a successful pilot transitions to a sustained production capability. The result is the problem the VP of Engineering encountered in the opening scenario: a working model becomes a production dependency without the governance, budget, or ownership structure needed to maintain it.

A stage-gate migration process creates the discipline needed to prevent this failure mode. Stage gates are decision points, not milestones, where a structured review determines whether a system is ready to advance to the next level of scale and operational commitment. They create explicit moments for the organization to ask, "Is this system performing as expected?" Is the team ready to own it at the next level of scale? Is the governance infrastructure in place? Does the business case still hold?

A practical stage-gate framework for AI systems has four stages: Discovery, Pilot, Limited Production, and Full Production. Each transition requires a formal review against defined criteria and produces a documented decision. The review is not a rubber stamp; it is a genuine gate, capable of holding a system at its current stage, requiring remediation before advancement, or, in some cases, recommending that the initiative be discontinued.

Stage one: Discovery is the period of problem validation and technical feasibility assessment. The goal is not to build a model. It is to determine whether the problem is AI-appropriate, whether the data exists to support an AI approach, and whether the business case justifies the investment. The Discovery stage-gate review should produce one of three outcomes: proceed to pilot, return for additional problem definition, or discontinue.

Stage two: Pilot is the period of model development and initial validation. The pilot operates on a bounded data set and a limited user population. Success criteria are defined before the pilot begins, not after. The Pilot stage-gate review assesses whether the model meets its technical performance criteria, whether the pilot user population has validated that the model's outputs are useful in the workflow context, and whether the data governance requirements for a limited production deployment have been met.

Stage three: Limited Production is a controlled deployment to a broader user population, with monitoring in place and a defined escalation process. This stage is when many organizations make their first serious investment in model-monitoring infrastructure, because the user population and business stakes are high enough to make drift a real risk. The Limited Production stage-gate review assesses model stability over time, monitors effectiveness, evaluates the team's capacity to manage the system at full production scale, and assesses budget alignment for recurring maintenance.

Stage four, Full Production, is the sustained operational state. At this stage, the model has a defined owner, a recurring maintenance budget, a weekly model review cadence, and a monthly impact review cadence. It appears in the organization's AI system inventory and is subject to periodic compliance reviews. The transition to Full Production is not the end of the governance process; it is the beginning of the sustained governance commitment that keeps the model performing and compliant over time.

Diagram 1.4 – Stage-Gate Migration Flow

AI Altalinora Infogram

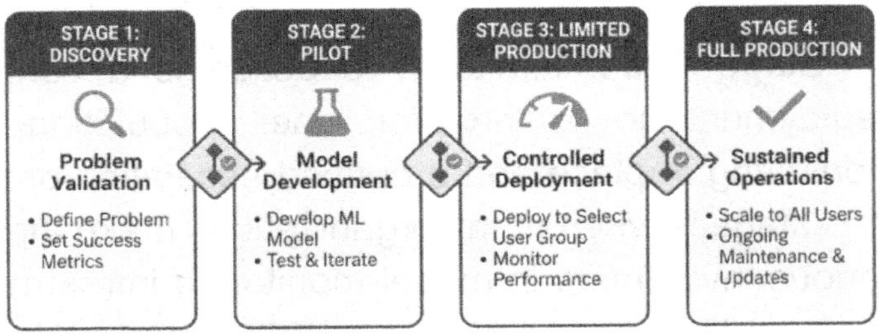

AI systems advance through four stage gates, each requiring explicit approval before proceeding.

1.7 Governance and Funding

Governance is the set of mechanisms that ensure AI systems are built and operated responsibly, consistently with the organization's values, and in compliance with applicable regulations. Funding is the mechanism that ensures AI systems have the resources they need to be maintained over their operational life. The two are more closely linked than they appear: governance without funding is a compliance exercise; funding without governance is a risk accumulation engine. Effective AI operating models address both simultaneously.

Budgeting for Recurring Model Maintenance

The most common budgeting failure in AI programs is treating model development as a capital investment and model maintenance as discretionary. This framing is dangerously wrong. An unmaintained model will degrade. Degraded models produce worse outputs. Worse outputs pose business, regulatory, and user trust risks. Model maintenance, including monitoring, retraining, data refreshes, and periodic governance reviews, is an operational necessity, not a discretionary activity.

A practical maintenance budget formula starts with the cost per inference, the compute cost of running the model against a unit of input, and works forward to a total operating cost estimate based on projected inference volume. It then adds the human costs: the AI Product Owner's time spent on monthly impact reviews, the data scientist's time spent on quarterly model evaluations, and the Data Steward's time spent on data refresh and governance compliance activities. For most models in production, annual maintenance costs run between fifteen and thirty percent of the original development cost.

Organizations that fund AI programs at the project level, allocating budget to build a model and then assuming it will operate indefinitely without additional investment, systematically underfund maintenance. The result is a portfolio of degrading models. The correction is structural: AI systems in Full Production should have a dedicated maintenance budget line that is reviewed annually and adjusted for changes in inference volume and data refresh requirements.

For organizations operating in cost-constrained environments, a common reality in public sector and nonprofit contexts, the maintenance budget question is a forcing function for portfolio discipline. If you cannot fund the maintenance of a model, you should not build it. A smaller portfolio of well-maintained models delivers more value and less risk than a larger portfolio of degrading ones. This is an argument for strong governance at the Discovery and Pilot stages, not for avoiding investment in AI.

Approval Thresholds and Risk Tiers

Not every AI system poses the same level of risk, and governance mechanisms should be calibrated accordingly. A recommendation engine that suggests blog content to a marketing team poses a different governance challenge than a clinical decision support system that suggests treatment protocols to physicians. Applying identical approval thresholds to both wastes governance capacity on low-risk systems and creates a false equivalence, making it harder to focus attention where it is genuinely needed.

A risk tier framework assigns each AI system to a tier based on the potential consequences of its failures. Tier-one systems are advisory tools with low stakes and an easy human override for content recommendations, search ranking adjustments, or internal productivity tools. Tier two systems influence consequential decisions, but with human review, a risk flag in an underwriting workflow, a patient outreach prioritization tool, and a budget allocation model. Tier three systems make or directly enable consequential decisions with limited human review, such as autonomous transaction processing, clinical protocol suggestions in high-volume care settings, and benefits eligibility determinations.

Approval thresholds should be calibrated to the tier. Tier-one systems may proceed through stage gates with standard technical review and product-owner sign-off. Tier two systems require additional stakeholder review, documented bias assessment, and explicit sign-off from the Data Steward and a relevant domain expert. Tier three systems require all the above, plus legal or compliance review, executive sponsor sign-off, and a formal risk assessment. The RACI for releases should explicitly reflect these tier distinctions, so that every person involved in a release decision understands their role.

Risk tiers should be assigned at the Discovery stage and reviewed at each subsequent stage gate. A system that begins as Tier one may migrate to Tier two or three as its scope expands. This recommendation engine, which began as an internal tool, may become a customer-facing decision system with much higher stakes. The governance process should include an explicit check for tier migration at each stage gate.

Health Metrics

A governance framework without health metrics cannot demonstrate its own value. Health metrics for an AI operating model fall into two categories: system-level metrics that measure the performance and reliability of individual AI systems, and portfolio-level metrics that assess the health of the AI program.

Time-to-value is the portfolio-level metric that the VP in the opening scenario cared most about: how long it takes from initial Discovery approval to measurable business impact. A mature operating model should reduce time-to-value over successive cohorts of AI initiatives as teams develop reusable infrastructure, clearer governance processes, and accumulate domain knowledge. If time-to-value is not improving, it signals that the operating model is not functioning as intended.

The percentage of models in production with active monitoring is a governance health metric that indicates how much of your AI portfolio is actively managed. An organization that has deployed forty models but monitors only fifteen has a governance gap of twenty-five model systems that could be drifting, failing, or accumulating compliance risk without anyone noticing. This metric should be tracked at the portfolio level and reviewed monthly. The target is 100% for Tier two and Tier three systems, and at least 80% for Tier one systems.

Additional health metrics worth tracking at the portfolio level include: the percentage of AI initiatives that pass their Discovery stage gate on the first review (a measure of problem definition quality), the average time spent in each stage (to identify systematic bottlenecks), the rate of model incidents per deployed system per quarter (to track governance effectiveness over time), and the ratio of maintenance budget to development budget across the portfolio (to track whether the organization is under-investing in maintenance relative to development).

Diagram 1.5 – Governance and Funding Dashboard

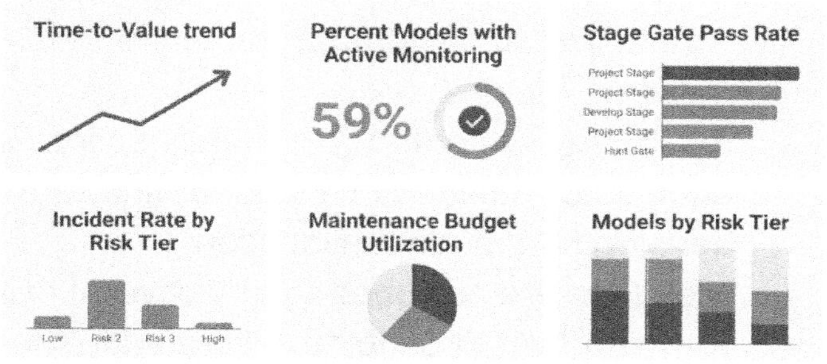

Track six health metrics to demonstrate AI governance effectiveness.

1.8 Manager's Checklist

Use this checklist to assess your current AI operating model and identify the highest-priority gaps to address.

- You have selected an operating pattern, centralized, hybrid, or distributed, and can articulate the criteria that led to that choice.
- Role charters for AI Product Owner, Platform Lead, and Data Steward are documented and assigned to named individuals.
- A stage-gate process exists, with defined criteria for each gate and a documented decision-maker for each transition.

- Every AI system in production has a designated AI Product Owner and appears in a maintained system inventory.
- A weekly model review cadence exists for each production system, with monitoring dashboards accessible to the review participants.
- A monthly impact review cadence exists, connecting AI system performance to business outcome metrics.
- Every AI system in Full Production has a dedicated maintenance budget line, not just a development budget allocation.
- Risk tiers are assigned to all active AI systems, with approval thresholds calibrated to the tier.
- The Operating Model Canvas, Stage-Gate Form, and Model Card Pointer artifacts exist for each active initiative.
- Portfolio-level health metrics, time-to-value, monitoring coverage, and incident rate are tracked and reviewed monthly.
- A postmortem process exists for AI system incidents, with findings documented and corrective actions tracked.
- The team has reviewed the operating pattern against the pattern that will be needed at the next-stage scale and has a migration plan.

1.9 Key Takeaway

The most expensive operating model failure is not choosing the wrong pattern. It is not choosing any pattern at all. Organizations that allow AI work to accumulate without a defined operating structure pay the cost in multiple currencies: technical debt, governance risk, talent friction, and, most expensively, the erosion of organizational trust in AI as a capability. Every pilot that fails because of missing ownership, every model that degrades because of absent monitoring, and every compliance finding that a documented approval process could have prevented are costs paid for the absence of an operating model decision.

The goal is operational clarity. Every person who touches an AI system- building it, using it, reviewing it, funding it- should understand what they own, what they own, and what success looks like. That clarity does not require a massive bureaucracy. It requires three chartered roles, three running cadences, three maintained artifacts, and a stage-gate process that creates deliberate decision points at every major milestone. Start there. Refine as you scale. Migrate to a more advanced operating pattern when the evidence, not the ambition, supports it. That is the pilot-to-production pathway that delivers workflow-level impact without the chaos that can end careers and erode board-level confidence in AI investment.

2 Designing Workforce Transformation That Sticks

2.1 Scenario From Skeptics to Advocates

The Director of Member Services at a regional insurance carrier had a problem she had not anticipated. Her team had been selected to pilot an AI-assisted claims triage tool, a system that would route incoming claims to the right specialist based on automated complexity scoring. On paper, it was an efficiency win. In practice, by the end of week two, she was fielding daily complaints. Claims specialists felt the system was making their expertise invisible. Two of her most experienced adjusters were actively routing around it. One submitted a formal objection through HR. The Director has not done anything wrong on the technological side. The model performed as designed. What she had missed was the workforce side: no one had explained why the tool existed, what it would and would not do, or how the specialists' domain knowledge was embedded in its design. No one had been trained in how to interpret the scores it produced. No one had been given a meaningful role in validating and improving it. Six months later, after a deliberate rotation program paired senior specialists with the data team and embedded a micro-learning curriculum in the team's weekly workflow, the same team was producing internal case studies on how to use the tool more effectively. The most vocal skeptic had become the team's informal AI champion. The tool's utilization rate increased from 41% to 93%. Nothing about the model changed. Everything about the people strategy did.

2.2 Why Workforce Transformation Is More Than Training

The most common mistake organizations make in AI-enabled workforce transformation is treating it as a training problem. Deploy the tool, run a training session, and assume the workforce will adapt. This framing misunderstands what transformation requires. Training addresses knowledge gaps. Transformation addresses identity, authority, trust, and incentive gaps, which are almost always more significant.

When a workflow is redesigned around an AI system, the people in that workflow experience changes in what their expertise means and in which decisions they are authorized to make. An experienced claims adjuster who has spent fifteen years developing judgment about claim complexity does not experience an AI triage tool as a productivity improvement. They experience it as a challenge to the organizational value of that judgment. That experience is not irrational, and a forty-five-minute onboarding session will not resolve it.

Effective workforce transformation addresses four distinct dimensions simultaneously: knowledge (can people use the tool competently?), authority (are people clear about what decisions remain theirs?), trust (do people believe the tool is reliable and fair?), and incentives (does the performance management system reward the new way of working, or the old?). An intervention that addresses only one of these dimensions typically will fail to produce the behavioral change that workflow-level impact requires.

For the enterprise IT and product leader, workforce transformation is not a soft topic. It is the critical path variable that determines whether a technically successful AI deployment translates into a business outcome. The health system that builds the best readmission prediction model in the industry but cannot get clinical staff to act on its outputs has not delivered value. The insurer that deploys a sophisticated claims scoring system but cannot get its adjusters to trust and use it has not delivered value. Technical performance is necessary but not sufficient. Workforce transformation is what converts technical performance into business results.

This chapter provides a practical framework for workforce transformation that is designed for enterprise realities: constrained training budgets, heterogeneous workforce skill levels, political dynamics around role changes, and the pressure to show impact quickly while building capabilities that will sustain AI adoption over the years. Every tool in this chapter has been validated in contexts where something was at stake: a production system, a regulatory requirement, a quarterly business objective.

2.3 Building the AI Role Taxonomy

A role taxonomy is not an org chart. It is a map of the competencies, accountabilities, and decision rights a team needs to build, operate, and benefit from AI systems. Role taxonomies matter for two reasons. First, they make hiring conversations concrete rather than debating whether a role should be called a 'data scientist' or an 'ML engineer'; a taxonomy defines what the role does and where it sits in the workflow. Second, they make reskilling conversations concrete instead of asking a business analyst to 'learn AI '; a taxonomy identifies the specific competencies that an analyst needs to develop to fill an 'AI-literate domain expert' role.

In a mature AI-enabled organization, four roles appear consistently across teams and domains. They are not always distinct individuals in smaller organizations; one person may fill multiple roles, but the four sets of competencies and accountabilities they represent are always present, whether or not they are explicitly named. Making them explicit is the first step toward building them deliberately.

AI Product Owner

The AI Product Owner is the role that bridges business strategy and AI capability. This person is accountable for defining what an AI system should accomplish, establishing the success criteria that will determine whether it has accomplished it, and making the product decisions about scope, prioritization, and trade-offs that govern its development and evolution. They are not the technical lead and should not be. Their job is to ensure that technical decisions serve business outcomes.

The competencies that distinguish an effective AI Product Owner from an effective traditional product owner are primarily in AI literacy. An AI Product Owner must understand the difference between a classification model and a regression model, not at a mathematical level, but at the level of knowing which types of business problems each is suited for. They must understand what training data requirements mean for the timeline and cost. They must be able to read a model performance report and ask the right questions about what the metrics mean for the specific use case. They must understand what model drift is and why it requires maintenance.

These competencies are learnable. An experienced product manager with a growth mindset and access to targeted AI literacy training can develop them in three to six months. What cannot be easily developed is the product intuition, the feel for user needs, business trade-offs, and organizational dynamics that make a product owner effective. The better hiring strategy is almost always to find an excellent product owner and invest in their AI literacy, rather than to find an AI expert and hope they develop product instincts.

ML Engineer

The ML Engineer is responsible for building, training, deploying, and maintaining the technical components of AI systems. This role sits at the intersection of software engineering and data science: they must be strong enough engineers to build production-grade infrastructure, and strong enough data scientists to understand model behavior and troubleshoot performance issues. In practice, many organizations discover that their data science teams have strong modeling skills but limited production engineering skills. The ability to build a good model in a notebook is not the same as deploying and monitoring it at scale.

The ML Engineer role is the one most organizations are actively hiring for. It is also the role where demand most consistently outpaces supply, which means organizations that rely entirely on external hiring to fill this role will consistently be talent-constrained. The more sustainable strategy is to develop ML engineering skills among existing software engineers, which is a shorter path than it may appear, since strong software engineers already have the production discipline that pure data scientists often lack and must develop.

For organizations in regulated industries, the ML Engineer role carries an additional compliance dimension. ML Engineers in healthcare, financial services, and federal contexts must understand the documentation and auditability requirements that apply to the systems they build. This is not purely a legal requirement; it is a professional responsibility. A model that cannot be explained, audited, or reproduced is a liability. ML Engineers who internalize that responsibility produce better systems than those who treat it as compliance overhead.

Data Steward

The Data Steward role is the one most organizations underinvest in and most deeply regret underinvesting in when a data governance issue surfaces. This person is accountable for the quality, provenance, privacy compliance, and representational integrity of the data that AI systems consume. In many organizations, data stewardship is distributed across multiple functions, including IT governance, privacy, legal, and data engineering, without any single owner. That distribution creates accountability gaps that AI makes much more consequential.

The Data Steward's key activities include: maintaining data dictionaries and lineage documentation, ensuring that training data complies with applicable privacy regulations, conducting representational audits to identify population gaps in training data that could lead to biased model behavior, managing data access controls and consent frameworks, and documenting data governance status in model cards for each AI system. This is a role that requires both technical literacy understanding data pipelines, schema management, and query tools) and legal and regulatory literacy in the specific domains that apply to the organization's data.

In organizations where the Data Steward role does not exist, the work falls to the ML Engineer by default, which is inappropriate. ML Engineers are built to optimize model performance. Data governance is a distinct discipline with distinct priorities, stakeholders, and risk frameworks. Giving data governance responsibility to the person whose primary metric is model accuracy creates a systematic conflict of interest. The Data Steward role exists precisely to manage that tension.

AI-Literate Domain Expert

The AI-literate domain expert is the fourth essential role and the one most often overlooked in org chart conversations because it does not always appear as a distinct job title. It is, instead, a competency layer that needs to exist within the business teams that will use, benefit from, or be affected by AI systems. A clinical informaticist who can evaluate a readmission risk model's output in the context of a specific patient population. A credit analyst who understands what a model score means and can identify when a score does not match their domain intuition. A claims adjuster who can articulate why a routing recommendation seems wrong and feed that feedback into a model improvement process.

AI-literate domain experts are the human layer that makes AI systems useful in practice. They are the people who catch edge cases, provide labeled corrections, generate feedback that drives model improvements, and build organizational trust in the systems they work with. The most reliable way to develop them is through deliberate exposure to AI development processes: inviting domain experts to model review sessions, including them in pilot cohorts, providing structured opportunities for model feedback, and recognizing their contributions in performance reviews.

The goal is not to turn domain experts into data scientists. It is to give them enough AI fluency to be effective partners in the AI development and operation process. That fluency is achievable through targeted, workflow-embedded learning, not through semester-long courses or generalized AI certifications that have no direct connection to the tools and systems the expert will work with.

Diagram 2.1 – AI Role Taxonomy Map

Four essential AI roles span technical and business dimensions across strategy and operations

2.4 Hire or Reskill: Making the Right Call

The hire-versus-reskill decision is one of the most consequential talent choices an AI-enabled organization makes, and it is rarely made with sufficient rigor. Organizations default to hiring because it feels faster to post a job description, run interviews, and make an offer. But in the current AI talent market, hiring is not fast for most organizations. Senior ML Engineers and experienced AI Product Owners are in high demand. Time-to-fill for these roles routinely runs six to nine months at organizations without elite employer brands. And even after the hire is made, onboarding, cultural integration, and the learning curve for domain context add another 3 to 6 months before the new hire is contributing at full effectiveness.

Reskilling takes time, too, but it has a different cost profile and a different set of risks. The primary cost is learning time, the portion of an existing employee's working hours dedicated to developing new skills. The primary risk is skill ceiling: some roles require competencies that cannot be developed through training alone, and some existing employees will not develop the target competencies even with significant investment.

The decision framework has three dimensions: speed to impact, cost to competence, and cultural fit. Speed-to-impact favors hiring when the skill gap is large and the timeline is short, but only if the hiring market can deliver. Cost-to-competency favors reskilling when the target competency is achievable from the existing base and the cost of training is significantly lower than the cost of a competitive hire, which is often the case for AI literacy and data stewardship roles, even when it is not the case for senior ML engineering roles. Cultural fit almost always favors reskilling: internal hires know the domain, the organization, the regulatory context, and the political dynamics that an external hire would need months to learn.

A practical decision rule: hire for roles where the technical skill requirement is the primary constraint and the domain knowledge requirement is low. Reskill for roles where the domain knowledge requirement is high, and the AI skill requirement can be reached from the existing competency base with a six-to-twelve-month investment. For most organizations, this means hiring ML Engineers externally while reskilling product managers into AI Product Owners and business analysts into AI-literate domain experts.

The financial case for reskilling is often stronger than it appears in initial budget discussions. A senior ML Engineer hire at a competitive salary may cost $200,000 per year in total compensation, $60,000 in recruiting costs, and 6 months of ramp time before full contribution. A reskilling investment of $20,000 per employee, spread across a cohort of 10 business analysts developing AI literacy, produces 10 AI-literate domain experts who already know the domain, the systems, and the regulatory context. The ROI calculation is not even close in most enterprise contexts.

Diagram 2.2 – Hire vs. Reskill Decision Framework

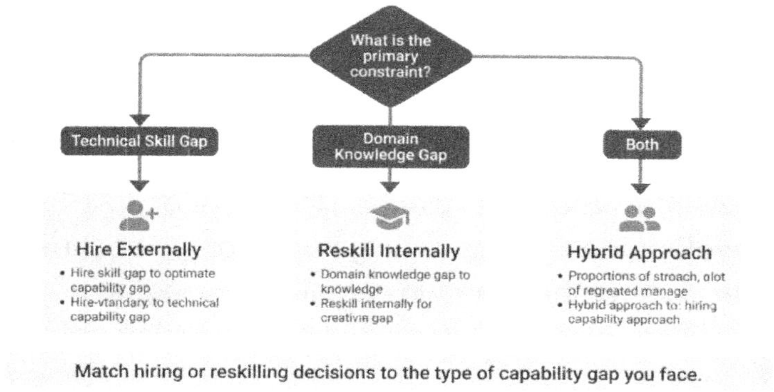

2.5 Practical Learning Pathways

Workforce transformation is not a single learning event. It is a sustained process of skill acquisition, practice, and reinforcement embedded in the actual workflow context where the skills are needed. Learning pathways are the structured sequence of learning experiences that move an individual from their current competency level to a target competency level over a defined time horizon.

Effective learning pathways for AI are fundamentally different from traditional training programs in one key respect: they are organized around doing rather than knowing. The goal is not for employees to describe how a machine learning model works. It is so they can make better decisions in their workflows thanks to AI fluency. That capability is built through practice in context, not by watching videos or attending lectures, however well designed.

Micro-Learning Modules

Micro-learning modules are short, focused learning experiences, typically five to fifteen minutes, designed to deliver one specific, actionable competency. They are the most scalable format for AI literacy development across a large workforce, because they can be embedded in existing workflows without requiring extended blocks of time away from work responsibilities.

The design principle for effective AI micro-learning is immediacy: every module should connect directly to a task the learner will perform in the next one to three days. A module on interpreting model confidence scores should be deployed the week before a new AI-assisted workflow goes live for that learner's team. A module on submitting model feedback should be deployed in the same sprint as the feedback interface. Learning that is immediately applicable has dramatically higher retention and behavior-change rates than learning delivered weeks or months before the relevant task context appears.

Micro-learning modules work best as a supplement to, not a replacement for, other learning formats. They build foundational literacy and reinforce specific skills.

They do not develop the contextual judgment and collaborative competencies that come from project-based learning and rotation experiences. The learning architecture should use micro-learning for breadth and speed, and cohort projects and rotations for depth and judgment.

Cohort-Based Projects

Cohort-based projects bring small groups of learners, typically four to eight people from different roles and domains, together to work on a real AI problem over a structured eight- to twelve-week learning arc. The problem is real in the sense that it matters to the organization, and its output will be used: genuine data exploration, a pilot model evaluation, a workflow redesign exercise. Learning is project-based in that the skills being developed- data literacy, model interpretation, product thinking, and domain integration- are practiced on real problems, not on synthetic exercises.

The cohort model has a powerful secondary effect beyond skill development: it builds cross-functional relationships and organizational trust in AI. When a clinical informaticist, a data scientist, and a care manager work together on a readmission risk project, they develop shared language, mutual respect for expertise, and a realistic understanding of what AI can and cannot do in that context. Those relationships and that shared understanding are organizational assets that outlast any single project.

For cohort projects to produce learning that sticks, they need three design features. First, they must have a defined scope that is achievable within the learning arc. Overscoped projects that do not reach a meaningful conclusion are frustrating for participants and produce less learning than tightly scoped projects that deliver a tangible result. Second, they must have a learning facilitator who is distinct from the technical lead, someone whose job is to help participants extract learning from their project experience, not just to ensure the technical work is done correctly. Third, they must have a showcase moment, a presentation of findings to senior leadership or a cross-team audience that gives participants recognition for their learning and creates organizational visibility for what was accomplished.

Embedded Learning in Workflows

Embedded learning is the most underutilized lever in enterprise workforce transformation. The idea is simple: the best time to learn a skill is in the moment when you need to perform the task that requires it. Rather than scheduling training events separate from the workflow, embedded learning inserts just-in-time prompts, contextual guidance, and in-product learning experiences directly into the tools and workflows where AI is being used.

A model inference interface that includes a 'Why did the model score this?' link, leading to a two-paragraph explanation of the features driving that specific score, teaches model interpretation to every user who clicks it. A feedback mechanism that prompts users to flag model outputs they disagree with and provides a brief, structured explanation of why they do so, creating a feedback loop that develops model literacy while simultaneously improving the model. A dashboard that shows a user how their team's utilization of an AI tool correlates with their productivity metrics provides behavioral reinforcement that accelerates adoption.

Embedded learning requires upfront design investment at the product level. It means the AI Product Owner and the product designer need to think about learning as a feature, not an afterthought. Organizations that embed learning into their AI products achieve higher adoption rates, higher user satisfaction scores, and faster development of AI-literate domain experts than those that rely exclusively on separate training programs. The investment pays back in both human capital and product utilization.

Diagram 2.3 – Learning Pathway Architecture

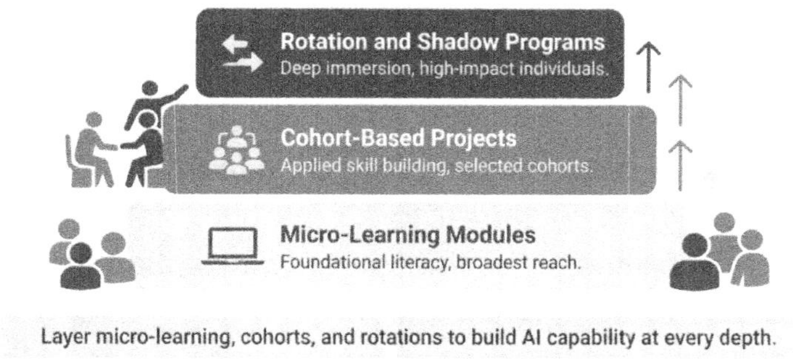

2.6 Rotation and Shadow Programs

Rotation and shadow programs are the highest-investment and the highest-impact workforce development mechanisms available to AI-enabled organizations. They work because they create genuine role immersion, a period in which a participant experiences AI work from the inside, not as an observer but as a contributing team member. The skills, relationships, and organizational perspective developed during a rotation cannot be replicated by any training program, no matter how well-designed.

A shadow program is a lighter-weight first step. A domain expert spends two to four weeks working alongside an AI team, attending design sessions, sprint ceremonies, model review meetings, and data discussions. They are not responsible for technical deliverables but are encouraged to contribute domain context, ask questions, and document their observations. Shadow programs are particularly effective for developing AI-literate domain experts because they demystify the AI development process without requiring participants to acquire technical skills they may not need.

A rotation program is a deeper commitment: a participant moves from their home team to an AI team or vice versa for three to six months, taking on real responsibilities. A business analyst rotating into an AI product team might own the backlog for a specific model feature, lead user research sessions with the domain team, and manage the documentation for a stage-gate review. A data scientist rotating into a clinical operations team might shadow care managers for two weeks to develop domain understanding, then use that understanding to improve feature selection for a care gap model.

The organizational infrastructure for rotation programs is often underestimated, as two-way rotations, where both the home team and the receiving team benefit, require careful capacity planning. Managers must be willing to release high-performing employees for three to six months, which requires organizational incentives that reward developing talent for the broader organization, not just retaining it within a team. Performance review systems must credit participation in rotation, both for the rotating employee and for the manager who supported it. Without these incentives, rotation programs stall because individual teams rationally protect their best people.

The organizations that sustain effective rotation programs treat internal mobility as a competitive advantage rather than a talent risk. They have found that employees who rotate across AI and business teams develop organizational networks, domain breadth, and systems thinking that make them more effective leaders over time. Rotation alumni consistently report higher engagement, stronger cross-functional relationships, and greater confidence in their ability to contribute to AI initiatives than peers who did not rotate.

Diagram 2.4 – Rotation Program Flow

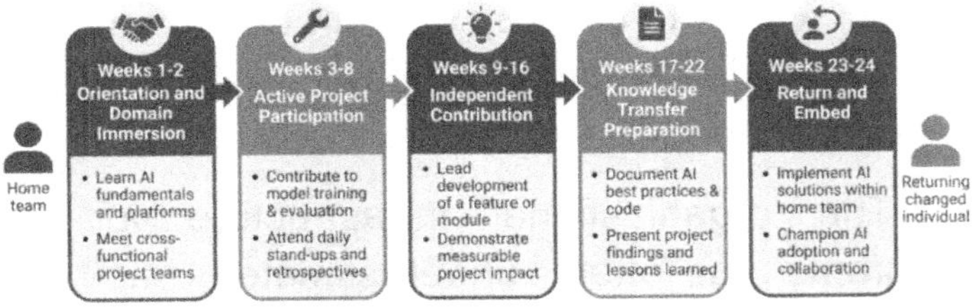

A structured six-month rotation transforms domain experts into confident AI collaborators.

2.7 Measurement and Incentives

Workforce transformation programs that do not measure their impact do not last. Budget cycles return, priorities shift, and without evidence of impact, learning and development investments are among the first to be cut. Measurement also serves as a more immediate management function: it tells you whether the program is working, where it is failing, and what adjustments are needed before those failures compound into larger organizational problems.

Measurement for AI workforce transformation operates at three levels: learning outcomes (did people acquire the target competencies?), behavioral change (are people applying those competencies in their workflows?), and business impact (is the behavioral change producing measurable results?). Most organizations measure only the first-level completion rates and assessment scores. Few measure the second. Fewer still connect investment in learning to business outcomes. The organizations that do all three sustain workforce transformation programs across multiple budget cycles.

Tracking Skill Adoption

Skill adoption is a behavioral measure, not a completion measure. An employee who completes a micro-learning module on model feedback has learned something. An employee who subsequently submits five feedback annotations in the AI system has adopted the skill. Completion metrics are easy to collect and tell you almost nothing about behavioral change. Skill adoption metrics require connecting learning records to system usage data, a more complex data integration, but a vastly more informative signal.

Practical skill adoption metrics to track include: rate of model feedback submission per AI-literate domain expert per week (measures model literacy in practice), number of model review sessions attended per AI Product Owner per quarter (measures governance engagement), ratio of domain experts who flag model outputs to domain experts who have completed the relevant training (measures training-to-practice conversion), and distribution of feedback quality scores over time (measures improvement in feedback sophistication, not just volume).

Tracking these metrics requires that AI systems and learning platforms share data. This is a product requirement, not an analytics afterthought. If the ability to connect learning records to system usage is not built into the product architecture from the start, it will be extremely difficult to add later. The AI Product Owner should include learning metrics and data infrastructure in the definition of done for every AI system with a learning pathway.

Incentive Design for AI Fluency

Incentives are the most powerful and most neglected lever in workforce transformation. It is easy to add AI fluency as a value or a stated organizational priority. It is both hard and essential to make AI fluency visible in the performance management system in a way that influences decisions about promotion, compensation, and recognition. Until it does, most employees will rationally allocate their development time to the competencies that are visibly rewarded.

The change in performance review language is the minimum intervention. Adding a specific competency dimension for AI fluency, with defined behavioral anchors at each performance level, makes AI skill development visible and comparable across the team. A senior analyst rated 'Developing' in AI fluency knows there is headroom for improvement. A manager reviewing their team sees AI fluency as an explicit dimension of professional growth rather than a nice-to-have. This single structural change in the performance review framework has measurable effects on voluntary participation in AI learning programs.

Promotion criteria should be updated to reflect AI fluency as a pathway accelerator rather than a prerequisite. Tying promotion to completion of AI certification creates bureaucratic friction and resentment. Recognizing demonstrated AI fluency through rotation participation, cohort project leadership, consistent model feedback contribution, or cross-team AI education delivery as evidence of the cross-functional impact that many organizations already include in senior-level promotion criteria creates a positive pull toward AI skill development rather than a compliance mandate.

Recognition programs outside the formal performance-review cycle provide additional behavioral signals. Internal AI challenge showcases, written case studies featuring teams that achieved workflow-level impact through AI adoption, and informal peer recognition in team meetings all contribute to an organizational narrative that treats AI fluency as a legitimate and valued form of professional excellence. These programs cost little and produce outsized cultural effects when run consistently.

Capacity Planning for HITL and Annotation

Human-in-the-loop, or HITL, and annotation work are among the most systematically underplanned workforce requirements in AI programs. The need for humans to review model outputs, label data, correct errors, and make decisions that the model is not authorized to make represents a real and ongoing labor commitment that must be incorporated into capacity planning from the earliest stages of AI program design.

HITL staffing requirements depend on three variables: the volume of model outputs that require human review, the decision time per review, and the quality standard the review must meet. For a fraud detection model that flags 300 transactions per day for human review, with an average review time of 4 minutes per transaction, the daily HITL labor requirement is 20 staff-hours, the equivalent of 2.5 full-time equivalents, assuming an 8-hour workday. This is a workforce planning input that must be incorporated in the business case before the model is built, not discovered after deployment.

Annotation budgets are a related planning challenge. Training a new model or retraining a drifted model requires labeled data, which in turn requires human annotation effort. Annotation is skilled work: a poorly designed annotation process produces noisy labels that degrade model quality. The annotation budget must account for annotator training, annotation time, inter-annotator agreement review, and quality auditing. Organizations that treat annotation as a commodity task and underpay for it get what they pay for in model quality.

The capacity planning process should include an annual HITL and annotation workforce review that examines: current staffing levels against current review volumes, projected volume growth and its staffing implications, quality metrics for current HITL review (are reviewers catching what they should be catching?), and automation opportunities that could reduce HITL requirements without compromising quality. This review should be a standard input to the annual AI program budget planning cycle.

Diagram 2.5 – Skill Adoption Measurement Dashboard

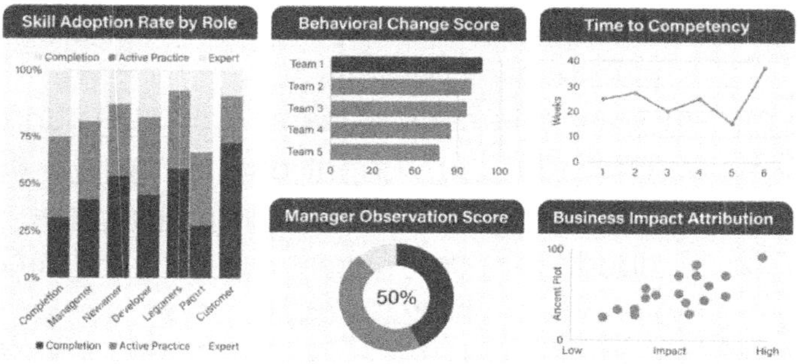

Measure skill adoption at three levels to prove and sustain workforce transformation.

2.8 Manager's Checklist

Use this checklist to assess the maturity of your workforce transformation strategy and identify the highest-priority gaps.

- Your AI role taxonomy is documented, with defined competencies, accountabilities, and decision rights for AI Product Owner, ML Engineer, Data Steward, and AI-Literate Domain Expert.

- Each active AI initiative has named individuals filling the four core roles, even if one person fills multiple roles in smaller teams.
- A hire-versus-reskill decision framework exists and has been applied to current talent gaps, with outcomes documented.
- A micro-learning curriculum exists for each major AI system in production, with modules deployed in the sprint of the workflow launch rather than weeks after.
- At least one cohort-based project is running or planned for the current quarter, with a defined scope, learning facilitator, and showcase event.
- Embedded learning features, in-context guidance, model explanation links, and feedback mechanisms are included in the product requirements for all new AI systems.
- A rotation or shadow program exists with defined intake, hosting, and return processes, and managers are incentivized to participate.
- Skill adoption metrics, not just completion metrics, are tracked for active learning programs and reviewed quarterly.
- Performance review language includes a defined AI fluency competency dimension with behavioral anchors at each performance level.
- Promotion criteria have been reviewed and updated to recognize demonstrated AI fluency as a pathway accelerator.

- HITL staffing requirements are calculated and incorporated into the workforce plan for every AI system in production or approaching production.
- An annotation budget is included in the business case for any AI system that requires labeled training data for development or ongoing retraining.

2.9 Key Takeaway

Workforce transformation that sticks is not a training program. It is a sustained, multi-dimensional intervention that addresses knowledge, authority, trust, and incentives simultaneously and measures its impact in terms of behavioral change and business outcomes rather than completion rates. The organizations that get this right are the ones that treat workforce development as a product problem: designing learning pathways with the same rigor they apply to AI system design, instrumenting those pathways to generate feedback data, and iterating based on evidence rather than intuition.

The lesson of the opening scenario is that the technology was never variable. The claims triage model performed as designed from day one. What changed between 41% and 93% utilization was not the model; it was the people working alongside it: their understanding, trust, authority, and incentives. Building those human conditions is the primary job of every leader responsible for AI-enabled workforce transformation. Technical excellence is the baseline. Workforce transformation is the differentiator.

3 Build a Culture That Enables Safe Experimentation

3.1 Scenario: The Demo That Changed Everything

Marcus had been CIO at a mid-size regional insurer for three years when he noticed a pattern. His AI pilots were technically solid. The models worked. The infrastructure was held. But adoption was slow, failure was silent, and experimentation was happening only where he personally applied pressure. When he asked his senior managers why, the answers were uncomfortable: people were afraid to be associated with something that did not work. The culture rewarded certainty. Being visibly involved in an AI experiment that failed was a career risk, not a learning opportunity. Marcus decided to change the signal. He started a weekly twenty-minute session, which he called the Thursday Lab, open to anyone in the technology and operations organizations. The format was simple: one team presented something they were testing, including what had not worked. He presented first, sharing a model he had personally sponsored that had failed to outperform the existing rule-based system. He talked about what he had learned and what the team would try next. He asked questions, not to evaluate, but to understand. Within six weeks, the Thursday Lab had a waiting list of teams wanting to present. Within three months, it became the organization's primary mechanism for spreading AI knowledge laterally across teams. The cultural change was not due to a training program or a policy update. It was a ritual that made failure visible, made learning valued, and made the CIO's behavior the clearest possible signal of what was safe and what was rewarded.

3.2 Why Culture Is the Operating System for AI Adoption

Strategy sets direction. Operating models set the structure. Culture sets the conditions under which everything else either functions or fails. An organization can have the best AI strategy in its sector, the most elegant operating model, and the most skilled data science team, and still fail to achieve AI-enabled impact if the culture punishes the experimentation that AI development requires.

AI development is inherently experimental. Models are not built; they are discovered. A team might invest twelve weeks in a modeling approach only to find that the data does not support the hypothesis. A pilot that works beautifully on historical data might underperform in production because the live data has different characteristics from the training set. A tool that makes ten percent of the workflow more efficient might simultaneously create friction in another twenty percent that was not anticipated. These are not failures of competence. They are the normal part of the discovery process in AI development, and in cultures where they are treated as failures of competence, they will be hidden rather than shared.

Hidden failures are among the most expensive organizational dynamics in AI programs. When a team discovers that their model is underperforming but does not report it for fear of the consequences, the organization loses the signal it needs to course-correct. Other teams continue to invest in similar approaches. The governance mechanisms designed to catch exactly this kind of drift have no information to act on. The underlying problem compounds until it is too large to hide, at which point the cost of remediation is significantly higher than it would have been if the failure had been surfaced and addressed immediately.

Culture change is difficult to produce through mandate. A memo declaring that the organization values learning from failure will not change behavior if the performance management system punishes failure and rewards only success. Culture is produced by the aggregate of visible behaviors, especially those of leaders with organizational authority. When a CIO shares their own failed experiment in front of two hundred employees, the signal that is carried is unambiguous: failure in service of learning is safe here. No policy document can produce that signal.

This chapter provides a practical framework for the cultural dimension of AI adoption. It is organized around concrete, manageable actions that enterprise leaders can take, not an inspirational aspiration to create the conditions in which safe experimentation is the default rather than the exception. Every element of this framework is designed to be implemented by managers who have limited control over formal culture change initiatives but significant influence over their teams' behavioral environment.

3.3 Creating Psychological Safety for AI Experiments

Psychological safety is the organizational condition in which team members believe they can speak up, take risks, and make mistakes without fear of punishment or humiliation. The concept, developed by organizational psychologist Amy Edmondson and extensively validated in subsequent research, is the single most consistent predictor of team learning effectiveness. Teams with high psychological safety learn faster, surface problems earlier, and produce better outcomes in complex, uncertain environments, which accurately describes the context of AI development.

Psychological safety is not the same as comfort. A psychologically safe team is not one in which all standards are relaxed, and all behaviors are accepted. It is a team where the discomfort of taking an intellectual risk, proposing an unconventional approach, or admitting a mistake is not compounded by the social risk of being judged, dismissed, or penalized. The distinction matters because managers sometimes resist the concept of psychological safety. After all, it sounds like an invitation to lower standards. It is the opposite: high-performing teams require psychological safety precisely because high performance in complex domains requires learning, and learning requires the willingness to be wrong in front of others.

Leader Behaviors That Build Trust

Psychological safety is built through repeated, consistent behavioral signals from leaders. It is not built through a single all-hands speech or a values poster on the wall. The behavioral patterns that matter most are: modeling intellectual humility, responding to failure with curiosity rather than judgment, explicitly inviting challenge and disagreement, and visibly protecting people who raise uncomfortable findings.

Modeling intellectual humility means that leaders acknowledge the limits of their own knowledge and share their own learning experiences, including their failures. This is difficult for leaders in organizations where authority is partly derived from expertise. A CIO who built their career on knowing more than the people around them may find it culturally costly to admit that an AI initiative they sponsored did not work as expected. That cost is real, and it is the price of the cultural shift. The leaders who pay it consistently find that it is far lower than the cost of the alternative: an organization where failure is hidden, and learning is suppressed.

Responding to failure with curiosity means that when an AI experiment does not produce the expected result, the first organizational response is a question: 'What did we learn?' rather than an evaluation: 'Who made the mistake?' This behavioral pattern must be consistent to be credible. A leader who asks, 'What did we learn?' nine times and then responds to the tenth failure with 'Who let this happen?' has undone the nine. Consistency is the only mechanism through which behavioral signals become cultural expectations.

Explicitly inviting challenge and disagreement in AI review sessions is particularly important for producing the diverse perspectives that improve AI quality. A model review that focuses solely on the model owner's interpretation of performance data will miss stakeholder concerns, domain-specific edge cases, and workflow friction points that would have been visible if others in the room felt safe raising them. The practice of actively asking 'What am I missing?' or 'Who has a concern about this that hasn't been voiced?' is not a formality; it is the mechanism that surfaces the information that prevents bad deployments.

Visibly protecting people who raise uncomfortable findings is the most powerful and rarest leadership behavior in this set. When a data scientist flags that a model has a performance gap in a specific demographic segment, when a product manager raises concerns about a deployment timeline, or when a domain expert says that the model's outputs do not match their professional judgment, the leader's response to those moments is the most visible cultural signal in the organization. A leader who thanks the person publicly, takes the concern seriously, and demonstrates that raising it was the right thing to do creates a cultural expectation that will propagate far beyond that individual conversation.

Failure Postmortems That Teach

The postmortem is an organizational ritual with enormous cultural leverage. Run well, it is a mechanism for converting organizational failures into organizational learning. Run poorly, it is a mechanism for distributing blame and reinforcing the fear of failure. The difference is almost entirely in the framing and facilitation.

A blame-free postmortem is not a postmortem without accountability. It is a postmortem where the analytical frame is systemic rather than individual. The questions are: what conditions made this failure possible? What processes failed to catch it? What could be changed in the system, workflow, governance process, monitoring infrastructure, or team structure to reduce the likelihood of this failure recurring? When these questions are answered well, the postmortem produces specific, actionable improvements. When the question is 'who made the mistake?', the postmortem yields specific, reluctant admissions and a team that is slightly more afraid of being in one the next time.

A postmortem for an AI system failure should examine the failure at multiple levels: the technical level (what did the model do that was unexpected?), the process level (what governance steps should have caught this before it caused impact?), the communication level (was the right information reaching the right people before the failure escalated?), and the structural level (were the ownership and accountability arrangements clear enough to enable a rapid response?). Each level of analysis produces a distinct category of corrective action.

The documentation habit turns postmortems from one-time events into organizational learning assets. A well-structured postmortem document covering the timeline of the failure, the contributing factors at each level, the corrective actions with owners and deadlines, and the governance process changes being implemented is a reference artifact that new team members can learn from, that governance reviewers can examine, and that leadership can use to track whether the organization is improving its ability to prevent and respond to AI system failures over time. A library of fifty well-documented postmortems is one of the most valuable organizational knowledge assets an AI-enabled enterprise can own.

Diagram 3.1 – Psychological Safety Framework

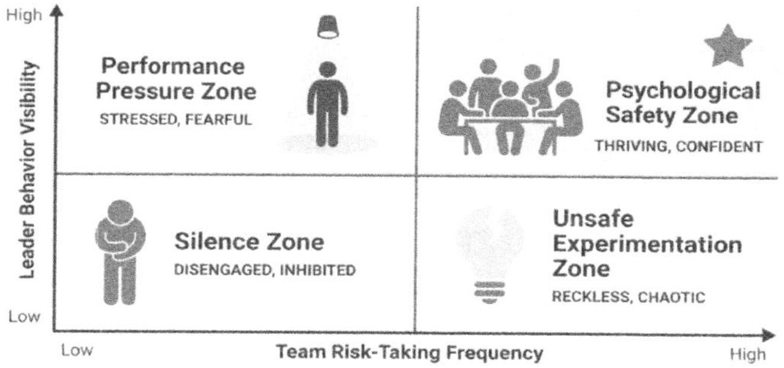

High leader visibility plus high team risk-taking defines the psychological safety zone.

3.4 Incentives and Recognition

Incentive structures are the infrastructure of organizational culture. They are the mechanisms through which an organization's formal values, stated in strategy documents and all-hands presentations, become behavioral reality or organizational hypocrisy. An organization that says it values learning from failure but whose performance review system rates only outcomes, not learning, will produce a workforce that prioritizes safe, predictable activities over risky, potentially valuable experiments. The gap between stated values and incentive structures is one of the most common root causes of AI adoption failure.

Closing that gap does not require a wholesale redesign of the performance management system. It requires targeted additions and modifications that make AI experimentation visible, valued, and rewarded in the formal recognition infrastructure that employees already navigate. The principle is: make the behaviors you want to see legible to the people you want to see them in, within the reward systems they already care about.

Learning KPIs

A Learning KPI is a performance indicator measured at the team or program level that directly tracks organizational learning behaviors rather than just outcomes. Examples include the number of postmortems completed per quarter (measures learning infrastructure), the number of experiments designed with pre-specified success criteria (measures intellectual discipline in experimentation), the ratio of experiments with documented learnings to total experiments (measures learning documentation discipline), and cross-team knowledge sharing events per quarter (measures lateral learning diffusion).

Learning KPIs matter for two reasons. First, they create a measurement infrastructure for the behaviors that drive long-term success of AI programs, which are often invisible in outcome-only measurement systems. A team that ran 12 experiments and documented lessons from all 12 is producing value, even if none of them reached production. Without learning KPI, that team's contribution is invisible. Second, what gets measured gets managed: teams accountable for learning KPIs build learning habits into their workflows rather than treating documentation and reflection as optional activities they will do when they have time.

Learning KPIs should be set at the program or organizational unit level, not at the individual level. Measuring individual learning rates creates gaming behavior, in which people complete training modules to hit completion targets rather than to develop genuine capability. Measuring team and program-level learning outcomes fosters shared accountability and incentivizes genuine learning behavior.

Recognition Programs

Formal recognition programs for AI experimentation and learning are a high-leverage, low-cost cultural intervention. They work because they create public, organizational-level signals that specific behaviors are valued signals that individual manager feedback cannot produce at scale.

The most effective recognition programs for AI cultures specifically recognize learning, not just success. A 'Best Pivot' award, given to the team that most effectively changed course based on experimental findings, sends a different and more powerful signal than a 'Best Innovation' award given to the team with the most successful deployment. Both awards have value. But in an organization where success is already recognized, the gap in the recognition infrastructure is almost always on the learning side.

Internal case studies are among the most durable and high-leverage recognition mechanisms available to AI leaders. A one-to-two page written account of how a team used experimentation to solve a business problem, including the approaches that did not work and what was learned from them, produces three distinct value streams simultaneously: it recognizes the team, it shares knowledge with the broader organization, and it demonstrates to the organization that this kind of work is worth writing about and reading. Published in an internal knowledge base and linked in leadership communications, case studies become part of the organization's memory, shaping how future teams approach similar problems.

Senior leader involvement in recognition events amplifies their cultural effect. When a CIO attends the quarterly AI showcase and asks substantive questions about the experiments being presented, including those that did not work, the organizational signal is clear and unmistakable. Senior leaders' time is the scarcest resource in most organizations. Where they choose to spend it is where the organization's attention follows.

Diagram 3.2 – Incentive and Recognition Architecture

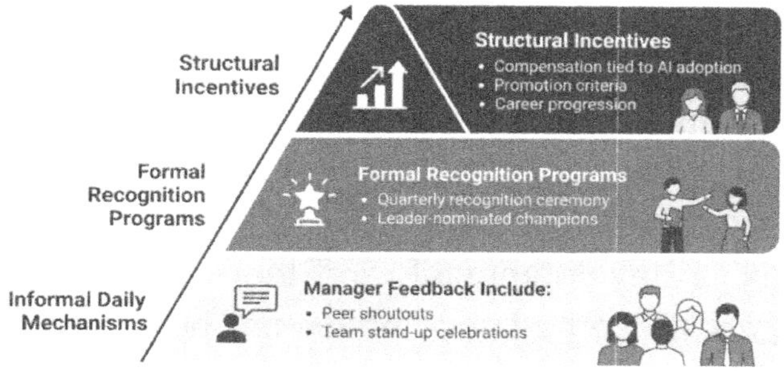

Layer informal recognition, formal programs, and structural incentives to sustain AI culture.

3.5 Rituals That Accelerate Learning

Organizational rituals are the regularly recurring practices through which a culture expresses and reinforces its values. In AI-enabled organizations, the rituals that matter most are those that create regular, structured opportunities for knowledge to flow across team boundaries, distributing the learning that accumulates in individual experiments across the wider organization before it becomes siloed, stale, or lost.

The word ritual is used deliberately. A ritual is not just a meeting. It is a recurring practice that carries meaning beyond its instrumental purpose. The Thursday Lab that Marcus ran was not primarily a knowledge-transfer mechanism; it was a ritual that signaled what the organization valued, who was safe to speak honestly to, and what the CIO thought was worth his time. Those signals were as important as any specific piece of knowledge exchanged in the sessions.

Demo Days

A demo day is a structured showcase event where AI teams present work-in-progress, not polished, production-ready deliverables, to a cross-organizational audience. The format is informal, the work is developmental, and the explicit norm is that what is presented need not be finished or successful to be worth sharing. What matters is that it is real and honest, and that the presenter is willing to discuss both what is working and what is not.

Demo days are most effective when they are held at a regular cadence, monthly or bi-monthly, and when the audience includes both technical and business participants. The cross-functional audience serves two purposes: it provides domain feedback to teams who are building things that will eventually need to work in business contexts. It develops AI literacy in the business audience through repeated, informal exposure to how AI works.

The design of demo days' facilitation matters enormously. A demo day that functions as a presentation event, where teams present and the audience observes, produces less value than one that functions as a conversation event, where teams present briefly and spend most of the time in structured discussion about their findings, challenges, and questions. The conversation is where the lateral learning happens: where team A discovers that team B encountered and solved the same data quality problem, where a business stakeholder flags a domain concern that the technical team had not considered, and where a junior engineer asks the question that reframes how the entire room is thinking about a modeling challenge.

Red-Team Reviews

A red-team review is a structured adversarial evaluation of an AI system or experimental design, conducted by a group whose explicit mandate is to identify flaws, risks, and failure modes that the development team may have missed. The red team is not an enemy of the project; it is the organizational mechanism that improves it by applying scrutiny that the team cannot apply to itself.

Red-team reviews are most valuable at two points in the AI development lifecycle: before a model is deployed to a pilot population (to identify technical and governance risks that should be resolved before users are affected), and before a model is promoted from pilot to production (to identify edge cases, bias concerns, and workflow integration issues that did not surface in the controlled pilot environment). They should be conducted by a diverse group that includes, at a minimum, a technical reviewer who did not build the model, a domain expert who understands the workflow context, and a governance reviewer who can assess compliance and risk.

The cultural function of red-team reviews is as important as their technical function. They normalize adversarial thinking as a productive contribution, not a form of obstruction or criticism, but as a form of quality assurance that the development team should welcome and seek out. Organizations that run red-team reviews consistently find that teams begin to internalize the red-team mindset, designing their experiments with adversarial scrutiny built in from the start. This is the cultural shift that makes responsible-by-design a practical operating mode rather than a compliance slogan.

Cross-Team Showcases

Cross-team showcases are demo-day events scaled for a cross-departmental or cross-divisional audience. Where the demo day is a team-level ritual, the cross-team showcase is an organizational-level ritual: a quarterly or semi-annual event where the AI program's collective learning across all active initiatives is presented to a broad audience that includes business leadership, IT leadership, and domain teams that may not have direct involvement in any active AI initiative.

The cross-team showcase serves a distinctive organizational function: it makes the AI program's collective output visible to an audience that is not already participating. For domain teams that have not yet engaged with AI development, the showcase is the primary mechanism through which they develop a realistic understanding of what AI can do in their context, what teams that have gone through the AI development process have learned, and what it would take to start an initiative of their own. This is how pilot-to-program scaling happens: not through top-down mandate, but through the grassroots emergence of informed enthusiasm that cross-team showcases consistently generate.

The production quality of cross-team showcases should be deliberately modest. Organizations that invest heavily in polished showcase presentations find that the production investment crowds out the honest, developmental content that makes showcases valuable. A showcase where every team presents a success story is a marketing event, not a learning event. The production quality signal should be that we care about the thinking, not the theater.

Diagram 3.3 – Learning Ritual Cycle

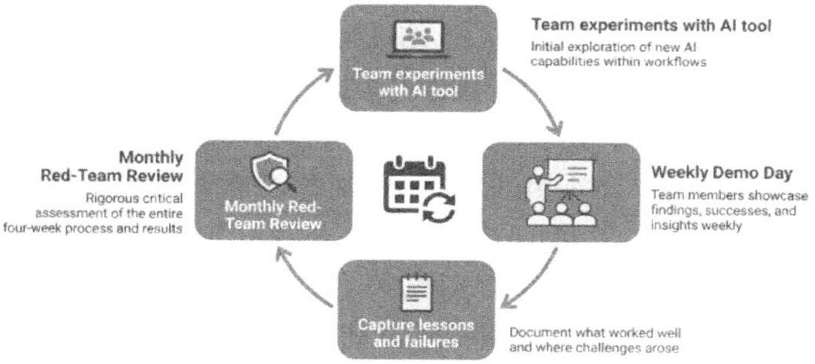

Weekly demos and monthly red-team reviews build continuous learning into team rituals.

3.6 Inclusion and Bias Mitigation as Cultural Practices

Inclusion and bias mitigation are sometimes framed as compliance requirements to be checked before a model can be deployed. That framing misses the most important insight: organizations that practice inclusion and bias mitigation as genuine cultural values produce better AI systems than organizations that treat them as regulatory overhead. This is not a moral argument, though the moral argument is valid as well. It is a quality argument. AI systems built by diverse teams with deliberate bias-mitigation practices make fewer costly mistakes than systems built without such practices.

The empirical basis for this claim is straightforward. Learn from AI systems' data. If the data they learn from is not representative of the full population the system will serve, they will perform worse for the underrepresented segments. If the teams building the systems do not include people with lived experience from the underrepresented segments, the gaps are less likely to be noticed and more likely to compound. If there is no organizational process for identifying and correcting these gaps before deployment, they reach users, with consequences that range from reduced effectiveness to active harm.

Embedding inclusion and bias mitigation as genuine cultural practices, not compliance formalities, requires making them visible at the team level, not just at the governance level. It requires specific rituals, roles, and processes to operationalize commitment to building AI systems that work fairly and effectively across the full range of people they will serve.

Inclusive Design Reviews

An inclusive design review is a structured evaluation of an AI system's design at the workflow, interaction, and model level for accessibility, representational adequacy, and cultural appropriateness. It is distinct from a bias audit, which is a technical evaluation of model performance across demographic segments. An inclusive design review occurs earlier in the development lifecycle, ideally at the Discovery and Pilot stages, when the cost of design changes is low.

A practical, inclusive design review examines four dimensions. First, accessibility: can the people who need to use or be affected by this system access it, given their language, literacy, technological access, and physical abilities? Second, representation: do the people who are building this system have sufficient direct knowledge of the experience of the people who will be most affected by it? Third, labeling: if the system uses human-labeled training data, are the labeling criteria clear and consistent, and have they been reviewed for implicit biases that could encode discriminatory patterns? Fourth, override: does the system design include clear, accessible, and dignified pathways for affected individuals to contest automated decisions or request human review?

Inclusive design reviews require participants who bring diverse perspectives. A review panel composed entirely of the technical team that built the system will not catch the blind spots most likely to cause problems. The review panel should include, at a minimum, a representative from the domain team who knows the population the system will serve, a representative from privacy or legal function, and, ideally, one or more representatives from the communities or populations most affected by the system's outputs.

Diverse Pilot Cohorts

The composition of a pilot cohort, the initial population on which a model is tested in production conditions, is a governance decision with direct consequences for model quality and organizational fairness. A homogeneous pilot cohort- a single team, a single location, and a single demographic profile will produce pilot performance data that does not accurately predict model performance across the full user population. The pilot may appear to succeed, leading to a full deployment that fails for segments not represented in the pilot.

Designing a diverse pilot cohort means deliberately including participants who represent the range of user experiences the model will encounter in production: varying levels of AI tool experience, different workflow contexts, different demographic profiles where relevant to model performance, and different organizational roles that interact with the system in different ways. This is not tokenism. It is quality engineering: producing pilot performance data that predict full-deployment performance.

The governance process should include a pilot cohort design review as part of the stage-gate approval for moving from Discovery to Pilot. The review question is: Is this cohort sufficiently diverse to yield meaningful performance data across the full range of use cases and populations that this system will serve? If the answer is no, the cohort should be redesigned before the pilot begins, not expanded post-hoc after a deployment reveals a performance gap in an underrepresented segment.

The Inclusive Pilot Checklist

An inclusive pilot checklist is a structured evaluation tool used during the Pilot stage to verify that a pilot has been designed and executed with adequate attention to representation, accessibility, and the risk of bias. It is a governance artifact, not a research instrument: its purpose is to create a documented, reviewable record of inclusion-related decisions made during the pilot, along with the evidence supporting them.

A practical inclusive pilot checklist covers: demographic representativeness of the pilot cohort relative to the intended deployment population, accessibility requirements that were verified during the pilot, documentation of any performance differentials observed across demographic segments during the pilot, description of the feedback mechanisms provided to pilot participants for surfacing concerns or contesting outputs, summary of any concerns raised by pilot participants and how they were addressed, and a signed attestation from the AI Product Owner and Data Steward that inclusion requirements have been reviewed and met.

The checklist is most effective when integrated into the stage-gate process as a required input to the Pilot-to-Limited Production gate, rather than maintained as a separate compliance activity. When inclusion review is embedded in the same governance infrastructure as technical quality review and business case validation, it is treated as an equivalent requirement, not as an afterthought that can be deferred when the timeline is under pressure.

Diagram 3.4 – Inclusive Pilot Design Flow

3.7 Moving from Small Wins to Scaled Rituals

Culture change begins with small, visible signals. But small signals alone do not produce organizational transformation at scale. The progression from a single leader's behavior to an organization-wide cultural norm requires deliberate infrastructure: structures, programs, and practices that sustain and amplify the initial cultural signal across organizational units, geographies, and leadership generations.

The risk of relying solely on individual leader behavior is that the culture becomes dependent on the personal commitment of one or two people. When those leaders move to new roles, the culture they built often does not survive the transition because it was shaped by their specific behavioral environment, not embedded in the organizational systems that persist through leadership changes. Scaling culture means making the behaviors, rituals, and norms structural rather than personal.

Leader Learning Sprints

A leader learning sprint is a structured, time-boxed learning experience designed specifically for senior leaders who need to develop sufficient AI fluency to be credible sponsors and effective decision-makers for AI initiatives. The format is distinct from the micro-learning and cohort project models designed for front-line employees: it is compressed, context-heavy, and geared toward people with limited time and high organizational stakes.

A leader learning sprint typically runs over four to six weeks, with two to four hours of engagement per week structured as a mix of curated reading, a facilitated cohort session with peers, direct exposure to an in-progress AI initiative (including a session with the technical team where the leader can ask naive questions without institutional consequences), and a reflective application exercise where the leader identifies how the learning applies to a current decision they are responsible for.

The most important design feature of a leader learning sprint is psychological safety within the cohort. Senior leaders are often the people in an organization who are least comfortable acknowledging the limits of their knowledge, because their authority is partly based on being perceived as knowledgeable. A leader learning sprint in which a VP of Operations admits they do not understand what model drift means, a high-value organizational event is, but only if the environment makes that admission safe. Cohorts of peers at similar levels, facilitated by an expert who models intellectual humility, create that environment.

The organizational effect of leader learning sprints extends far beyond the participants. When a VP returns from a sprint and begins asking different questions in AI review sessions- not 'when will this be ready?' but 'what does the model card say about data representativeness?' The signal to their team is immediate and unmistakable. The leader's behavioral change multiplies with every team meeting, every resource allocation decision, and every casual hallway conversation in which their newly developed AI fluency is evident.

Visible Sponsorship

Visible sponsorship is the sustained, active demonstration by a senior leader that an AI initiative has their personal attention and support. It is distinct from the organizational sponsorship shown on a project charter, as indicated by the name in the 'Executive Sponsor' box. Visible sponsorship means the sponsor shows up at demo days, stage-gate reviews, team retrospectives, and postmortem sessions. It means they speak about the initiative in external and internal contexts. It means they use their organizational authority to remove barriers when the team encounters them.

The cultural function of visible sponsorship is not primarily to authorize resources, though resource authorization is necessary. It is signal amplification. When employees across the organization see that a senior leader is personally engaged in an AI initiative, they draw accurate inferences about organizational priorities, career safety, and the seriousness of the organization's commitment. Those inferences shape behavior: people volunteer for AI work, participate in learning programs, and take the risks that experimentation requires when they believe the organization's most powerful actors have genuinely committed to this direction.

Visible sponsorship is also the primary mechanism for breaking organizational friction that technical teams cannot address on their own. An AI team that needs data access from a resistant IT governance function, a product team that needs capacity from a business unit that is protecting its headcount, a governance process that has stalled because no one has decision authority in each of these cases, a visible sponsor who makes a phone call or attends a meeting can resolve in an hour what would otherwise take months to untangle through formal channels.

Scaling Playbooks and Recognition

Scaling rituals and cultural practices require that the knowledge underlying them be codified, transferable, and accessible to teams and leaders who were not present for the initial cultural moments that gave rise to them. A playbook is the primary codification mechanism: a structured, practical guide to running a specific organizational practice, a demo day, a red-team review, or an inclusive pilot design process that a team with no prior experience can use to implement the practice with reasonable effectiveness.

AI program playbooks should be treated as living documents rather than finalized artifacts. The teams that maintain them should have a regular cadence, at a minimum annually, to review, update, and improve them based on accumulated experience. The most valuable updates come from postmortem findings, demo day feedback, and rotation alums whose experience across multiple teams gives them a comparative perspective on what works and what does not.

A playbook library is most effective when paired with a community of practice: a voluntary, cross-organizational network of people actively working on AI implementation who meet regularly to share experiences, troubleshoot challenges, and learn from each other. Communities of practice are self-sustaining in ways that formal programs often are not, because the value they provide is immediate and practical rather than future-oriented and aspirational. They require modest organizational infrastructure: a meeting cadence, a communication channel, and a rotating facilitation responsibility. They produce organizational learning at a rate that no formal training program can match.

Recognition programs that acknowledge teams for their contributions to the organizational playbook, for writing a new guide, for updating an existing one based on their experience, for presenting their approach at a cross-team showcase, and for creating an incentive for the documentation work that sustains the cultural infrastructure. The recognition does not need to be large. It needs to be visible, consistent, and tied specifically to the behavior being recognized: contributing to the shared organizational knowledge that makes every team's work more effective.

Diagram 3.5 – Culture Scaling Roadmap

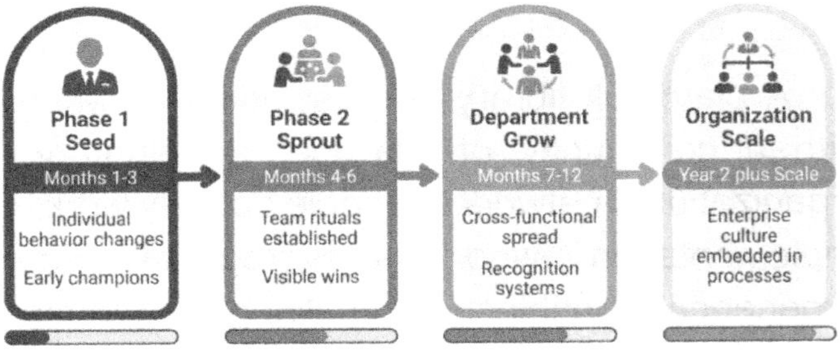

Culture scales in four phases from individual seed behaviors to enterprise-wide embedding.

3.8 Manager's Checklist

Use this checklist to assess your team and organization's cultural readiness for sustained AI experimentation.

- You can identify two or three specific behaviors you model that signal to your team that intellectual risk-taking is safe in your presence.
- Your team has a postmortem practice for AI experiments and model failures, with documented findings and tracked corrective actions.
- A regular demo day or equivalent learning ritual exists for your team, with a norm that work-in-progress and failures are shareable, not just successes.

- A red-team review process exists for AI systems before pilot deployment and before production promotion.
- Learning KPIs, not just outcome KPIs, are included in your team's performance measurement framework.
- Performance review language includes AI experimentation and learning behaviors as explicit competency dimensions.
- Your recognition practices include specific, public recognition for learning from failure, not just for successful deployment.
- Pilot cohorts for new AI initiatives are reviewed for demographic and contextual representativeness before the pilot begins.
- An inclusive pilot checklist is integrated into the stage-gate process for Pilot-to-Limited Production transitions.
- Inclusive design reviews are conducted at the Discovery and Pilot stages, with diverse review panels.
- At least one senior leader in your organization is visibly and actively sponsoring AI initiatives in the ways described in this chapter.
- A playbook library exists for at least the core AI rituals your team runs, and it is updated based on accumulated experience at least annually.
- A community of practice or equivalent cross-team learning network exists and meets on a regular cadence.

3.9 Key Takeaway

Culture is not the soft dimension of AI transformation. It is the operating system on which all technical and organizational capabilities run. The most sophisticated AI operating model, the most carefully designed workforce transformation program, and the most robust governance infrastructure will all underperform if they are running on a cultural substrate that makes failure dangerous, learning invisible, and experimentation irrational.

The good news for managers is that culture change does not require organizational authority to initiate. It requires behavioral consistency. A team leader who models intellectual humility, runs a blame-free postmortem after every AI system failure, and holds a fifteen-minute demo every other week is already changing the cultural environment for their team. When that behavior is sustained, visible, and critically protected and amplified by senior leaders, it propagates outward across organizational boundaries. The Thursday Lab that Marcus started with twenty people became an organization-wide ritual precisely because it was consistent, authentic, and visibly supported. That is the pilot-to-platform pathway for culture: start with a single, genuine behavioral signal, sustain it with structural reinforcement, and scale it with the deliberate infrastructure of playbooks, recognition, and community. The path from skeptics to advocates runs through exactly that progression.

4 Creating Learning Sys. That Convert Use into Skill

4.1 Scenario From Tool Usage to Team Mastery

The VP of Field Operations at a large regional health system had a problem she could not see in any dashboard. Her organization had deployed an AI-assisted documentation tool to roughly 2,400 nurses and care coordinators across fourteen hospitals. The vendor's license utilization reports showed a healthy 78 percent activation rate. Helpdesk tickets were low. Leadership was pleased. But when she started walking through units and asking specific questions, what did you learn from that output last Tuesday? How did you decide to override the recommendation? The answers were unsettling. Most staff were using the tool passively. They accepted the output without reading the reasoning behind it. They had learned the minimum needed to get through a shift, not the maximum that the tool was designed to enable. Skill accumulation had flatlined at the point of first competence. The tool was being used. It was not producing learning. She realized she was measuring adoption as a proxy for capability, and the two had stopped correlating. What she needed was not a training refresh. She needed a learning system, instrumented, structured, and tied to career progression that would convert daily workflow interactions into compounding skill.

4.2 Why Learning Systems Are the Missing Layer

Most enterprise AI programs invest heavily in two layers: the technology layer and the deployment layer. The technology layer covers model selection, infrastructure, security architecture, and integration. The deployment layer covers rollout planning, change management communications, and helpdesk readiness. Both are necessary. Neither is sufficient. What organizations consistently underinvest in is the learning layer, the set of instrumented, structured, feedback-driven mechanisms that convert daily tool usage into steadily improving human capability.

The gap is consequential. Without a learning layer, AI adoption follows a predictable decay curve. Early adopters gain skills rapidly because they are intrinsically motivated to experiment. Mainstream users plateau at minimum viable competency because there is no structured path forward. Laggards remain stuck at surface-level usage indefinitely. The result is a workforce that checks the adoption box on a deployment scorecard while leaving most of the value the AI system was designed to create unrealized.

For managers accountable for workflow-level impact, this is not an abstract concern. It translates directly into metrics that matter, such as patient documentation quality, underwriting cycle times, citizen service resolution rates, and software delivery throughput. When a team uses a tool without deepening their understanding of how to use it well, those metrics improve modestly at first, then plateau or decline as over-reliance replaces judgment. The learning layer is the mechanism by which initial tool adoption becomes a sustained performance lift.

An AI learning system has three distinct components. The first is instrumentation: the telemetry, feedback capture, and experiment infrastructure that generates learning signals from actual workflow behavior. The second is architecture: the programs, cohorts, embedded mentors, and credential structures that organize learning into a progressive path. The third is governance: the privacy controls, artifact retention policies, and ROI frameworks that make the learning system trustworthy and accountable. Each component is addressed in detail in the sections that follow.

4.3 Instrumenting Workflows for Learning Signals

You cannot manage what you cannot measure, and most organizations measure the wrong things when it comes to AI learning. License activations, login frequency, and session duration measure access, not learning. To build a learning system that improves over time, you need instruments that capture the quality of human-AI interaction, the moments where users engaged deeply, questioned outputs, corrected mistakes, or demonstrated judgment that AI lacked. These signals indicate where learning is happening and where it has stalled.

Instrumenting workflows to capture learning signals requires coordination among three stakeholder groups: the product team responsible for the AI tool, the people operations team responsible for development and performance, and the IT team responsible for data infrastructure. Managers who fail to convene all three typically end up with data that is either technically rich but organizationally disconnected (product team acting alone) or organizationally meaningful but technically superficial (HR acting alone). The combination is what enables mission-aligned learning measurement.

Telemetry for Learning

Telemetry for learning is distinct from operational telemetry. Operational telemetry tracks whether the system is working, including latency, uptime, error rates, and throughput. Learning telemetry tracks whether the human is improving override rates, edit distances, confidence calibration, query sophistication, and time-on-task trends. Each of these signals contains information that operational metrics do not.

Override rates measure how often users reject or modify AI outputs. A declining override rate in a stable user cohort can mean one of two things: users are improving their queries, so outputs require less correction, or users have stopped engaging critically and are accepting outputs uncritically. Distinguishing between these two interpretations requires cross-referencing override rates with downstream outcome quality, for example, whether documentation accepted without override produced better or worse patient outcomes than edited documentation. This kind of cross-referenced telemetry is what separates a learning system from a simple usage dashboard.

Edit distance measures the gap between an AI-generated output and the final human-approved version. Large edit distances early in a user's tenure are normal. What matters is the trajectory: does edit distance decrease over time (suggesting the user is learning to prompt more effectively), does it hold steady (suggesting the user is using the tool transactionally), or does it increase (suggesting the user is becoming more skeptical or the model is drifting from the user's context)? Tracking edit distance trajectories by cohort reveals which user groups are on a learning curve and which are not.

Query sophistication is harder to measure but highly valuable. It tracks whether users' questions to AI systems become more specific, contextual, and multi-step over time. A care coordinator who starts by asking 'what should I document here' and progresses to 'compare this patient's presentation to the last three similar cases and flag any deviations from the care pathway' demonstrates meaningful growth in capability. Operationalizing a query sophistication metric requires an NLP classification layer on top of the query log. Still, the investment is justified for high-value roles where skill depth is directly tied to outcome quality.

Diagram 4.1 – Learning Signal Architecture

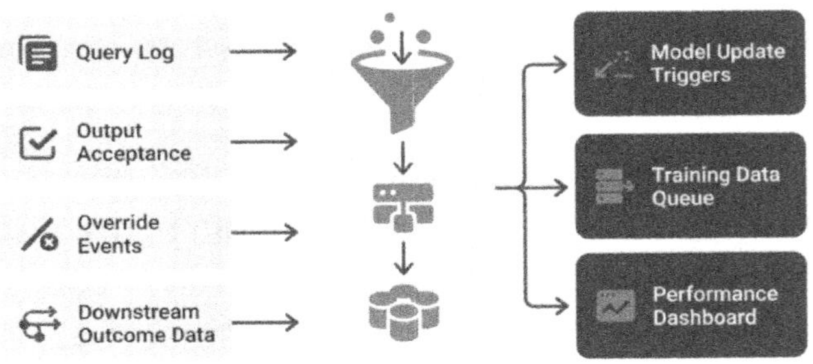

Workflow events flow through a telemetry pipeline to drive continuous model improvement.

Feedback Capture and Annotation Pipelines

Telemetry is passive data; it records users' actions. Feedback capture is active data; it records what users think. Both are necessary. Passive telemetry tells you the what; active feedback tells you the why. A user who consistently overrides a specific type of recommendation may do so because the model is wrong in that domain, because the user does not understand the recommendation, or because local policy diverges from the model's training data. Telemetry alone cannot distinguish between these causes. Structured feedback can.

Annotation pipelines for learning systems should be designed around three principles. First, capture as close to the moment of use as possible. Feedback requested immediately after a workflow decision is far more accurate and actionable than feedback collected in a weekly survey. Second, keep friction minimal. A binary thumbs-up/thumbs-down plus a single optional comment field captures 80 percent of the signal with 10 percent of the burden. Third, close the loop visibly. Users who never see their feedback acted upon stop providing it. Even a simple digest showing how feedback shaped a model update or a training adjustment signals that the feedback loop is real.

Annotation pipelines also support model improvement, not just human learning. When users correct AI outputs with contextual comments, they generate labeled training data that reflects your organization's specific operational context. A federal agency's underwriters, who regularly annotate why certain risk assessments diverge from model recommendations, produce a domain-specific signal that vendor models built on generic data cannot replicate. Building the infrastructure to capture, quality-check, and eventually feed this signal back into model fine-tuning is a significant competitive advantage. However, doing so responsibly requires governance and consent frameworks. These are addressed in the chapter's governance section.

Diagram 4.2 – Feedback Loop and Annotation Pipeline

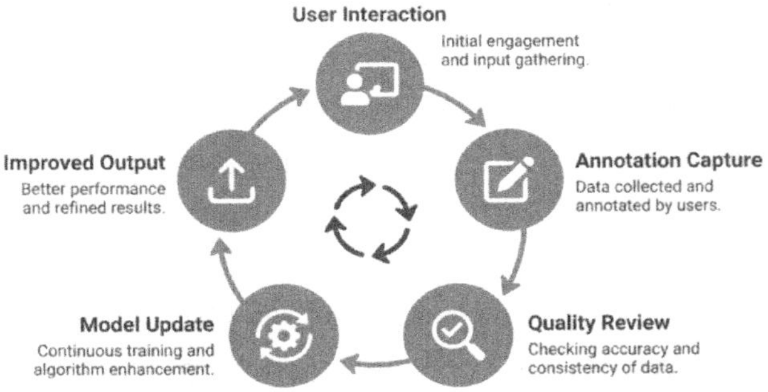

A continuous feedback loop from user interaction through annotation drives model improvement.

Experiment Design for Learning Transfer

Instrumenting a learning system is not the same as proving that it works. Proof requires experimental design and controlled comparisons that isolate the effect of specific learning interventions from the baseline effect of tool usage alone. Without this, you cannot distinguish between a learning program that genuinely accelerates skill development and one that merely appears to progress. At the same time, the real gains come from other sources.

The most tractable design for most enterprise environments is a controlled cohort experiment. Two user cohorts with similar baseline characteristics are given access to the same AI tool. One cohort receives the structured learning program: in-flow prompts, micro-credentials, embedded mentor check-ins, and annotated example libraries. The other receives standard onboarding. Both are measured on the same outcome metrics over a 90-day window. The difference in outcome trajectories provides a clean estimate of the learning program's marginal value.

Several design choices significantly affect the validity of cohort experiments in enterprise settings. Randomization should be at the team level, not the individual level, to avoid contamination if some members of a team are in the treatment group and others are in the control group; social learning will blur the distinction. Outcome metrics must be chosen before the experiment begins, not after, to avoid selection bias from choosing the metric that shows the best result. And the experiment window must be long enough to capture skill accumulation, which occurs over weeks to months, not days.

For managers at large health systems, insurers, or federal agencies, even a modestly sized cohort experiment with 50 users per arm provides sufficient statistical power to make credible investment decisions about the scale of learning programs. The cost of running the experiment is almost always less than the cost of scaling a learning program that does not actually work. Treat experiment design as a core management competency, not a research team luxury.

4.4 Building Practical Learning Architectures

Once you have instrumentation in place to generate learning signals, the next challenge is building the organizational structures that translate those signals into actual skill development. Signals alone do not produce learning. A user whose telemetry shows a plateau in query sophistication will not spontaneously begin asking better questions because the metric is declining on a dashboard. You need programs, social structures, and credentialing mechanisms that create both the opportunity and the motivation to improve.

Learning architecture for AI-enabled organizations differs from traditional corporate training in several important respects. They are embedded in workflows rather than separated from them. They are continuous rather than episodic. They are individualized by signal rather than scheduled by calendar. And they are tied to career outcomes, credentials, promotions, and role mobility rather than compliance requirements. Building an architecture with these characteristics requires deliberate design choices across three structural elements: AI academies, embedded mentors, and micro-credentials.

AI Academies and Cohort Design

An AI academy is a structured learning environment that brings together employee cohorts to develop AI-related capabilities through a combination of practice, peer learning, and guided instruction. The concept is modeled on cohort-based learning, as in executive MBA programs and high-performing technical bootcamp environments, where the social dynamics of a peer group learning together accelerate individual development beyond what solo study can achieve.

Effective AI academy design starts with cohort composition. Cohorts should be cross-functional rather than siloed by job family, because AI skill development is inherently interdisciplinary. A cohort comprising a process analyst, a data engineer, a product manager, and a frontline team leader develops a richer shared understanding than a cohort of 10 process analysts alone. The friction of explaining your workflow to someone who does not share your context is itself a learning mechanism; it forces articulation and reveals assumptions.

Cohort size should be constrained to enable genuine peer relationships. Research on learning cohorts consistently finds diminishing returns above 25-30 participants. For enterprise AI academies, cohorts of 12-20 members provide the right balance of diversity and intimacy. Running multiple concurrent cohorts is preferable to running a single large cohort.

Curriculum design for AI academies should follow a use-case-anchored model rather than a concepts-first model. Start with a real workflow challenge that cohort members are trying to solve, and introduce relevant AI concepts, tools, and techniques to address it. This approach produces higher retention and faster transfer because the learning is immediately applicable. A cohort working through how to use AI to reduce prior authorization processing time at a health insurer will retain far more about prompt engineering and output validation than a cohort that studies those topics in the abstract.

Diagram 4.3 – AI Academy Cohort Model

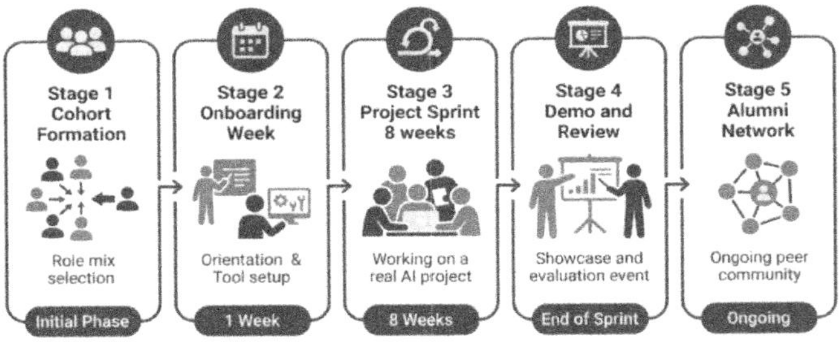

AI Academy cohorts move from formation through active project work to a lasting alumni network.

Embedded Mentors

Cohort programs build foundational capability. Embedded mentors sustain and deepen it. An embedded mentor is a senior practitioner, not a trainer or a vendor representative, who spends a defined portion of their time working alongside a team, available for in-flow questions, co-problem-solving, and the kind of just-in-time coaching that classroom instruction cannot replicate. In the context of AI-enabled work, the embedded mentor's most valuable contribution is modeling the judgment that separates expert use from novice use.

The distinction between an embedded mentor and a traditional subject matter expert resource is important. A subject matter expert is consulted when problems arise. An embedded mentor is present before problems arise, observing patterns in how the team uses AI tools, proactively surfacing where assumptions go unchallenged, and creating low-stakes opportunities to practice judgment in real workflow contexts. The value is not in answering questions; it is in helping the team develop the questions worth asking.

Embedded mentor programs require careful role design to avoid two common failure modes. The first failure mode is the mentor who becomes a resource router, fielding questions and directing people to documentation rather than doing actual coaching. This produces dependency rather than capability. The second failure mode is the mentor who becomes so integrated into the team's workflow that they become a bottleneck, the one person everyone consults before acting, rather than a catalyst for distributed judgment. Effective mentor program design includes explicit time limits (6-12 months per deployment), clear coaching objectives, and measurement of the mentor's impact on the team's independent capability rather than on the mentor's personal output.

For large field forces, the health system nurses, insurance adjusters, and government benefit processors, who number in the thousands, cannot be matched with embedded mentors on a 1:1 basis. A tiered model is more practical: embedded mentors within each major business unit, supported by a network of peer champions (often the early adopters identified during initial deployment) who handle first-level team-level coaching. This creates a learning infrastructure that scales without requiring a linear increase in headcount.

Micro-Credentials and Career Alignment

The single most powerful accelerant for workforce AI skill development is career alignment, which makes it clear that developing AI capabilities leads to concrete career outcomes. This is not about motivation in the abstract. It is about the very practical calculation every employee makes when deciding how to invest their discretionary learning time. If the answer to 'what happens to me if I get good at this' is 'nothing formal happens, but your manager will be pleased,' voluntary skill development will remain confined to the intrinsically motivated minority.

Micro-credentials are the mechanism that makes career alignment operational. A micro-credential is a verifiable, specific attestation of a defined skill or cluster of capabilities, for example, 'AI-Assisted Documentation: Advanced,' 'Workflow Automation Design: Practitioner,' or 'Model Output Validation: Specialist.' Unlike traditional training certificates, micro-credentials are tied to demonstrated performance rather than seat time, are specific enough to be meaningful to hiring managers and promotion committees, and are stackable into career paths.

Designing an effective micro-credential architecture requires answering three questions: What skills are being credentialed? What evidence demonstrates attainment? What organizational actions does the credential unlock? The third question is the most critical and the most often neglected. A credential that does not unlock anything- a promotion opportunity, a salary band adjustment, eligibility for a frontier role, or access to an advanced cohort program is just a badge. Badges without consequences do not change behavior at scale.

At the insurer deploying AI underwriting tools, a well-designed micro-credential architecture might look like this: Tier 1 credentials cover basic AI literacy and tool proficiency; they are required for continued use of AI tools after a six-month grace period. Tier 2 credentials cover workflow optimization and output validation; they are a requirement for promotion to senior adjusters. Tier 3 credentials cover system design and teaching; they are the pathway into the AI Center of Excellence. Each tier has defined assessment criteria, connects to a learning path that prepares candidates for the assessment, and is recognized in the compensation system.

Diagram 4.4 – Micro-Credential Career Ladder

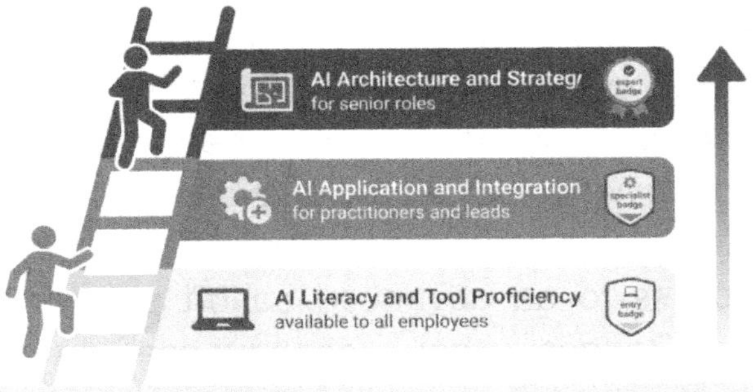

Three micro-credential tiers align AI skill recognition to career advancement pathways.

4.5 Mapping Learning Outcomes to Product KPIs

A learning system that cannot demonstrate its impact on product performance will not survive budget cycles. The accountability link between learning investments and business outcomes must be designed from the beginning, not retrofitted after programs are running. This requires a disciplined process of connecting learning outcomes to the skills people developed, at what pace, and to what level of mastery, to the product KPIs the AI system was deployed to improve.

The starting point for outcome mapping is the theory of impact: a written, falsifiable account of how skill development in a specific capability domain is expected to change a specific outcome metric. For example: 'As care coordinators develop greater proficiency in interpreting AI risk stratification outputs (learning outcome), the accuracy of their care escalation decisions will improve (proximate outcome), resulting in a reduction in preventable readmissions (product KPI).' This chain skill, proximating the outcome to the product KPI, is the backbone of an outcome mapping framework.

Once the theory of impact is established, the measurement architecture follows. You need measures at each link in the chain: a skill proficiency measure (tied to the micro-credential assessment or the telemetry signals), a proximate outcome measure (decision accuracy, as assessed by clinical audit), and a product KPI measure (readmission rate, tracked in the operational reporting system). All three measures need to be tracked at the same unit of analysis, ideally at the team level, so that variation within the chain can be attributed and addressed.

Several pitfalls in outcome mapping are worth naming explicitly. Attribution is always contested: readmission rates are affected by dozens of factors besides care coordinator decision accuracy. Address this by designing the measurement period and unit of analysis to control as many confounders as possible, and by being transparent about the confidence interval around the learning program's contribution estimate. Lag matters: skill development affects downstream outcomes on a delay that varies by domain but is rarely less than 60 days. Build lag assumptions into your measurement plan rather than discovering them after the fact when early results disappoint. And metric selection creates incentives: if the only metric you track is query volume, users will optimize for query volume rather than query quality. Choose metrics that reward the behaviors you want.

The operational clarity that comes from a well-designed outcome mapping framework is itself a management asset. When learning investments are tied to specific KPI lines, conversations about program funding shift from 'is this worth it in general' to 'what is the marginal cost of a one-point improvement in readmission rate, and does this learning intervention achieve that at competitive cost?' That is a conversation that can be won on merit.

4.6 Governance for Learning Systems

Learning systems generate and consume sensitive data about individual employees' performance, skill gaps, query patterns, and knowledge limitations. Governing this data responsibly is not optional, and in many jurisdictions, it is legally mandated. Beyond legal compliance, responsible governance of learning data is a prerequisite for employee trust. Workers who believe their learning signals are being used to surveil rather than support them will suppress the very behaviors that honest feedback, candid questions, and willingness to display uncertainty are needed for a learning system to function.

Governance for learning systems operates across three domains: privacy and data minimization, artifact retention and knowledge capture, and ROI metrics and accountability. Each domain has distinct requirements and distinct failure modes.

Privacy and Data Minimization

The privacy principle most relevant to learning systems is data minimization: collect only the data necessary to achieve the learning objective. This principle is both ethically sound and practically advantageous. Every additional data point collected beyond what is needed expands the attack surface for data breach, increases the burden of consent management, and creates the potential for data to be used in ways that erode employee trust.

Applying data minimization to learning telemetry requires answering a sharp question about each data element: what learning decision does this data point enable that we could not make with aggregated or anonymized data? Individual-level telemetry is often necessary for personalized coaching and competency assessment. Aggregate-level telemetry is sufficient for curriculum design and program evaluation. Drawing a clear line between these two uses and governing individual-level data with stricter controls are practical approaches to minimization in enterprise environments.

Consent for learning data collection should be informed, specific, and revocable. Employees should know what data is being collected, how it will be used, who will have access to it, and how they can request correction or deletion. Blanket consent buried in a system's terms of service agreement does not meet this standard and will not withstand either regulatory scrutiny or reputational challenge. In unionized environments, learning data governance will often be a subject of collective bargaining, and managers who treat it as a technical matter rather than an employee-relations matter will be caught off guard.

Role-based access controls on learning data should follow the principle of least privilege: each stakeholder role (employee, direct manager, HR business partner, learning system administrator) has access only to the data necessary for their legitimate function. An employee's direct manager can see their individual skill trajectory and coaching notes. The HR analytics team can see aggregate cohort trends but not individual identifiers. The AI product team can see anonymized telemetry. Access logs should be maintained and auditable.

Artifact Retention and Knowledge Capture

Learning systems produce artifacts that have value beyond their immediate use: annotated examples of high-quality AI output, records of complex decisions and the reasoning behind them, documented failure cases and lessons learned, and expert commentary on model limitations in specific operational contexts. Governing these artifacts- how they are versioned, stored, accessible, and eventually retired- is a knowledge management challenge as much as a data management challenge.

The highest-value artifacts in most enterprise learning systems are the decision rationale records produced by experienced practitioners as they annotate their own choices. A senior underwriter explains why she overrode an AI risk score for a particular account: it creates an artifact that contains domain expertise, model boundary knowledge, and policy context that cannot be reconstructed from raw telemetry. Retaining these artifacts, making them searchable, and incorporating them into onboarding materials for new hires creates a knowledge flywheel: the organization gets smarter over time, not just the individuals who happen to be working today.

Versioning matters for learning artifacts in ways that differ from standard document management. When the underlying AI model is updated or retrained, previously annotated examples may no longer accurately reflect the current model's behavior. Artifact governance systems need to track model versions alongside artifact versions. These flag artifacts may be stale relative to the current model and surface this staleness information to learning program administrators in time to update curricula before learners encounter misleading examples.

Retention policies for learning artifacts must balance knowledge preservation against privacy rights. Employees have legitimate interests in not having records of their early-career mistakes permanently accessible. A retention policy that archives individual-level learning artifacts for two to three years, then anonymizes them for program evaluation purposes, and permanently deletes identifiable data upon employee departure strikes a reasonable balance in most jurisdictions. Legal counsel should review applicable data protection regulations, particularly in the EU, where GDPR rights of erasure create specific obligations.

ROI Metrics for Learning Investments

The final governance requirement for learning systems is accountability for return on investment. This is not simply a financial reporting exercise; it is the mechanism by which learning programs are improved, resource allocation decisions are made, and organizational commitment is sustained. Without credible ROI measurement, learning investments are the first item cut when budgets tighten, regardless of their actual value.

ROI metrics for learning investments should operate at two levels: efficiency and impact. Efficiency metrics measure the cost and velocity of learning: cost per learner, time to achieve a defined competency milestone, completion rates, and learning program Net Promoter Score. These metrics are relatively easy to measure and provide the operational visibility needed to manage program delivery. They do not, however, answer the question that senior leaders care most about: what did the learning investment produce?

Impact metrics answer that question by connecting learning outcomes to business results. The strongest impact metrics connect directly to the product KPIs identified in the outcome mapping exercise. For a health system investing in care coordinator AI literacy, impact metrics might include average time-to-competency for new hires (compared to historical baseline), percentage of care escalation decisions rated accurate by clinical audit (compared to pre-program baseline), and contribution to readmission reduction target (estimated via the cohort comparison methodology described earlier). Each of these metrics has a defined numerator, denominator, measurement period, and comparison baseline.

The total cost of the learning investment should include all fully loaded costs: program design and development, facilitation, mentor time, technology infrastructure, and the opportunity cost of employee time spent on learning activities. Dividing total impact (expressed in a business unit the organization cares about, such as readmissions avoided, cycle time saved, or errors prevented) by total cost yields a cost-per-unit-of-impact figure that can be compared across programs and used to defend to finance. It will never be a perfect number. It is always better than no number at all.

Diagram 4.5 – Learning System ROI Dashboard

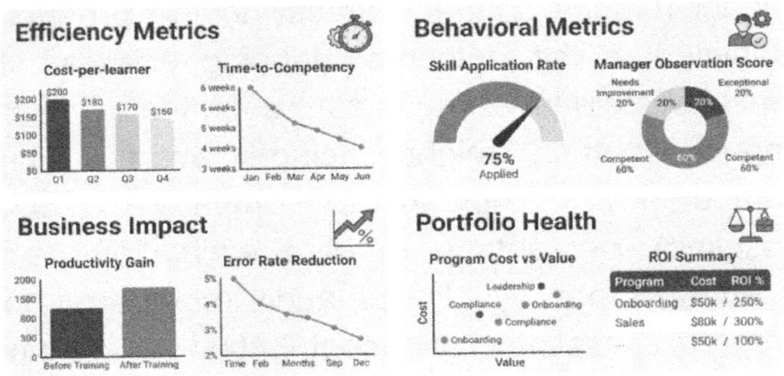

Measure learning system ROI across efficiency, behavior, business impact, and portfolio health.

4.7 Manager's Checklist

Use this checklist to assess the maturity of your organization's AI learning system and identify priority investment areas:

- Telemetry infrastructure: Are you capturing override rates, edit distances, and query sophistication by user cohort? Are these signals feeding a learning dashboard that is reviewed at least monthly?
- Feedback capture: Is in-flow feedback collection active in your AI tools? Is there a process for routing annotation data to both model improvement and learning program design?
- Experiment capability: Have you run at least one controlled cohort experiment to validate the marginal impact of a learning intervention? Is there a defined process for running future experiments?

- Academy program: Is there an active cohort-based learning program with cross-functional composition, use-case-anchored curriculum, and defined completion criteria?
- Embedded mentors: Is there a tiered mentor network covering all major business units that use AI tools? Are mentors measured by the independent capabilities of their teams, not by their personal output?
- Micro-credentials: Is there a published micro-credential architecture with defined assessment criteria and clear career consequences at each tier?
- Outcome mapping: Is there a documented theory of impact connecting each learning program to specific product KPIs? Are all links in the chain being measured?
- Privacy governance: Is data minimization applied to all learning telemetry? Is consent documented, specific, and revocable? Are role-based access controls current and audited?
- Artifact governance: Is there a versioning system that tracks learning artifacts against model versions? Is there a retention policy that legal counsel has reviewed?
- ROI accountability: Is there a defined ROI measurement framework covering both efficiency and impact metrics? Are results reviewed quarterly with senior leadership?

4.8 Key Takeaway

Tool deployment without a learning layer produces a temporary performance bump followed by a plateau. The organizations that extract sustained competitive and operational value from AI are those that treat learning as a system, not an event, one that is instrumented to generate signals, architected to develop skills, governed to protect trust, and measured to demonstrate impact. The VP of Field Operations, who recognized that her health system was measuring adoption instead of capability, had identified the right gap. Closing it required building all three components of the learning layer: the telemetry and feedback infrastructure, the academy and credential architecture, and the governance and ROI framework. The result, when all three components operate together, is a workforce whose capabilities compound over time rather than plateau at first use. That compounding capability is what separates organizations that use AI from those enabled by it.

5 Leading Change Programs for AI Adoption

5.1 Scenario: When the Skeptics Became Champions

The Director of Claims Operations at a mid-market property and casualty insurer had held the role for 11 years. She had seen every flavor of technology rollout, core system replacement, digital portal launch, and robotic process automation wave, and she had learned to read the warning signs early. When her CIO announced that an AI-assisted claims triage system would be deployed to her 340 adjusters over the following six months, her first reaction was not enthusiasm. Her adjusters were experienced professionals who took pride in the judgment they applied to complex claims. She anticipated that many of them would experience the new system as a threat, a slight, or at a minimum, an imposition. She was correct about the resistance. What she got wrong was the assumption that resistance was inevitable and permanent. She convened a co-creation lab: eight adjusters, including the most vocal skeptics, who spent four weeks working alongside the product team to design the adjuster experience, write the output-explanation templates, and build the decision-support interface that everyone else would use. Six months later, those eight adjusters had become the most credible advocates for the system in the organization. The program succeeded not because resistance was overcome through better communication, but because the skeptics became owners. That shift was the product of deliberate design, not accident.

5.2 Why AI Rollouts Are Change Journeys

AI adoption programs fail far more often due to organizational reasons than to technical ones. The model may be accurate. The integration may be stable. The security review may be complete. And yet the program produces no measurable change in operational performance because the humans who were supposed to use it have found ways to work around it, ignore it, or comply with it superficially without genuinely incorporating it into their decision-making. This pattern is consistent enough across industries and use cases to qualify as a near-universal law of enterprise AI: technology readiness does not equal adoption, and adoption does not equal value.

The reason is that AI systems, unlike most previous enterprise technologies, do not merely change how a task is performed. They challenge the basis on which professional judgment is exercised. A financial analyst who has spent a decade building intuition about creditworthiness is not simply being asked to use a new tool when an AI risk model is introduced; she is being asked to reconsider the epistemic authority of her own experience. A police dispatcher who has developed deep pattern recognition of neighborhood dynamics is being asked to defer, at least in part, to statistical inference. These are not neutral operational changes. They are identity-relevant disruptions and require the kind of deliberate change management that organizations typically reserve for mergers and restructurings, not for software deployments.

For managers accountable for the success of AI adoption programs, this reframing has practical implications. First, the timeline for genuine adoption- not activation, not compliance, but substantive behavioral change- is measured in quarters, not weeks. Planning that assumes a 30-day adoption window assumes a performance plateau at first-level compliance. Second, the stakeholders who matter most are not the early adopters, who will adopt regardless of program design, but the mainstream majority and the skeptics, who need different things from the program at different stages. Third, the measurement of success should track behavioral outcomes, such as decisions being made differently and better, not just usage metrics.

Leading a successful AI change program requires mastery of five interconnected disciplines: defining a compelling North Star that makes the destination legible; mapping stakeholders to understand the terrain; designing pilots as genuine experiments rather than rollout previews; measuring sentiment and adoption with the same rigor as technical metrics; and building champion networks that sustain momentum after the formal program ends. Each is addressed in the sections that follow.

5.3 Define the North Star

Every effective change program begins with a destination worth reaching, specific enough to be actionable, compelling enough to motivate, and durable enough to withstand the inevitable turbulence of implementation. For AI adoption programs, this destination is the North Star: a concise, verifiable statement of what the organization will be able to do when the program succeeds that it cannot do today.

The North Star is not a vision statement. Vision statements are intentionally aspirational and intentionally vague. Their function is to inspire, and vagueness protects them from being falsified by events. The North Star is intentionally specific and intentionally falsifiable. A weak North Star sounds like this: 'We will become an AI-enabled organization that leverages cutting-edge technology to deliver superior outcomes.' A strong North Star sounds like this: 'By the end of Q3, 90 percent of claims adjusters will be using AI triage recommendations as a first-pass input, and average claims resolution time will have decreased by 15 percent relative to the Q1 baseline.' The strong version can be measured. It can be decomposed into milestones. It creates accountability.

Vision Framing

Framing the North Star requires answering a set of questions that are deceptively simple and practically difficult. What will frontline workers be able to do after the program that they cannot do now? What will the system be able to do, but what will the humans using it be able to do? What operational metric will be demonstrably better, and by how much? What will the organization stop doing because AI makes it unnecessary? What new capabilities will the organization gain that were previously cost-prohibitive?

Vision framing also requires honesty about what the program will not deliver or will deliver only partially. AI adoption programs consistently overpromise and underdeliver when their sponsors frame the North Star as a total transformation rather than a specific operational improvement. Experienced managers and frontline workers are skilled at detecting inflated claims, and when they do, their skepticism extends to legitimate claims as well. A North Star that says 'this will change everything about how we work' will be believed by almost no one. A North Star that says 'this will take three hours per week off documentation burden for care coordinators, and those three hours will go to patient-facing activities' is believable, testable, and compelling to the people it directly affects.

The framing should be tailored to the audience. The North Star, as communicated to a CFO (cost efficiency, risk reduction, scalability), is not the same as the North Star as communicated to a frontline team lead (time savings, reduced repetitive burden, better information at the moment of decision) or to a regulator (consistent documentation, auditable decision trails, compliance efficiency). The underlying substance is the same; the framing emphasizes the dimensions most relevant to each audience's concerns. This is not spin; it is the basic work of connecting a complex change to the specific interests and fears of each affected group.

Success Metrics

Success metrics for an AI adoption program should be selected before the program begins, not after. Pre-registration of success metrics is the single most effective protection against the post-hoc rationalization that afflicts most enterprise change programs: the tendency to declare success on the metrics that happened to move favorably and to discount or explain away the metrics that did not.

A robust success-metric architecture for AI adoption comprises four layers. Adoption metrics measure whether users are engaging with the system as designed: activation rates, session frequency, feature utilization depth, and override rates. Proficiency metrics measure whether users are developing genuine capability: these are drawn from the learning system telemetry described in the previous chapter. Outcome metrics measure whether operational performance is improving cycle times, reducing error rates, improving decision quality, and increasing customer satisfaction. Organizational metrics measure whether the change program is building durable capability: internal-champion density, credential attainment rates, and knowledge-artifact production.

Each metric layer has a distinct time horizon. Adoption metrics move within weeks and provide early warning signals. Proficiency metrics move within months and validate the learning pathway. Outcome metrics vary by quarter and demonstrate business value. Organizational metrics over the years indicate whether the change is becoming self-sustaining. Reporting on all four layers simultaneously, at the cadence appropriate to each, gives leadership a complete picture of program health rather than the partial and sometimes misleading view that comes from tracking only one layer.

Diagram 5.1 – North Star Alignment Framework

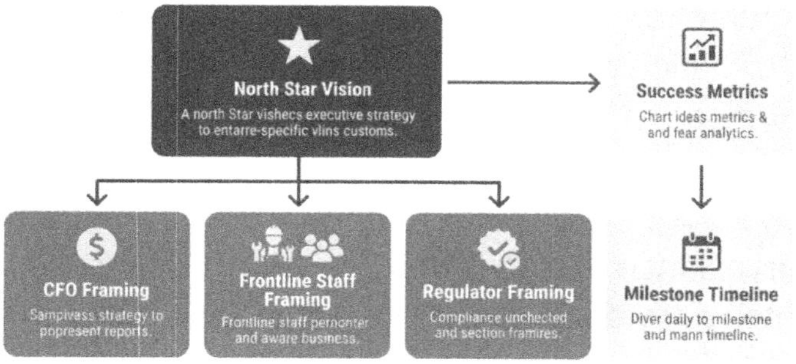

A North Star vision connects executive strategy to audience-specific framing and measurable milestones.

5.4 Stakeholder Mapping and Communication

Change programs fail when they treat stakeholders as a uniform audience. Every complex organization contains multiple stakeholder populations with different relationships to the AI system, risk profiles, sources of identity and professional pride, and needs from the change program. Effective stakeholder mapping makes this variation legible and actionable; it tells you who needs what, when, and through which channels.

The fundamental error in most stakeholder engagement for AI programs is the assumption that information is sufficient. Tell people enough about how the system works, why it was built, and the way the thinking goes, and resistance will dissolve. This assumption has been falsified repeatedly in practice. Information reduces uncertainty, but it does not change interests, allay identity threats, or address the very legitimate concerns that experienced practitioners have about deskilling, job displacement, performance surveillance, and the erosion of professional autonomy. Stakeholder engagement that addresses only the information deficit will not move the populations that matter most.

Identifying Early Adopters and Resistors

Stakeholder mapping for AI adoption begins with a three-segment taxonomy: early adopters, mainstream majority, and active resistors. Each segment requires a fundamentally different engagement strategy, and misapplying that strategy across segments is a common source of program failure.

Early adopters self-identify, request access before the formal rollout, experiment beyond the prescribed use cases, and become unofficial, informal advocates within their teams. Their primary need from the change program is not motivation; they are already motivated, but channel: access to information, early features, and influence over program design that rewards their energy without co-opting their independence. Early adopters who engage as co-designers and informal champions become the most credible voices within the organization. Early adopters who are left unengaged drift toward the informal role of impatient critics of the program's pace and conservatism.

The mainstream majority is the population whose adoption determines whether the program succeeds. They are not opposed to change; they are uncertain about whether this change is worth the investment of their trust and effort. They watch the early adopters closely, and their decision to engage is heavily influenced by what they observe. Programs that rely on direct persuasion of the mainstream majority, such as emails from the CIO or all-hands presentations, miss the social dynamics that drive their decisions. What persuades the mainstream majority is watching trusted peers use the system and report genuine benefit. This is why early-adopter activation and champion-network design are strategic, not tactical, priorities.

Active resistors are the smallest but loudest segment. They are often the most experienced and most respected members of their professional cohort, which is precisely why their resistance carries disproportionate weight. Their concerns typically center on substantive accuracy, erosion of professional judgment, and surveillance, and should be treated as such rather than as expressions of technophobia or obstructionism. A change program that dismisses active resistors' concerns will harden their resistance. A program that creates a genuine forum for their concerns, specifically including them in co-creation processes, often discovers that their skepticism identifies real product and governance gaps that improve the final program for everyone.

Diagram 5.2 – Stakeholder Mapping Matrix

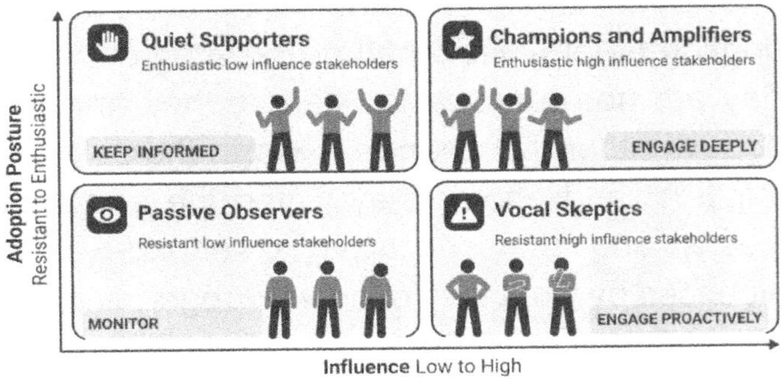

Map stakeholders by influence and adoption posture to tailor engagement strategies.

Designing Personalized Communications

Personalized communication in enterprise change programs does not mean sending individualized emails to every employee; it means segmenting communications by the concerns, motivations, and information needs that vary by role, tenure, and adoption posture. A single communication plan that tries to speak to everyone in the same voice will resonate with no one in particular.

For each major stakeholder segment, the communication plan should specify: what the core message is for this audience, what their primary concern is and how the message addresses it, which channel is most credible (executive email, team meeting, peer-to-peer conversation, visual demonstration), what the call to action is, and when the message is delivered relative to program milestones. The last element is often neglected. Sending a 'here is why this change matters' message after people have already encountered the system in their daily work, without prior context, produces a very different response than sending it when their questions are at their highest.

For senior leaders, communications should emphasize strategic rationale, risk management, and performance trajectories. For frontline teams, communications should emphasize day-to-day workflow impact, time savings, and the career development benefits of building AI proficiency. For middle managers, who often receive the least intentional communication and carry the heaviest implementation burden, communications should emphasize the support infrastructure available to them, the expectations that are and are not changing, and the recognition they will receive for leading successful adoption within their teams. Middle-manager communications are a chronic underinvestment in AI change programs and a chronic source of adoption failure.

5.5 Pilot Governance and Co-Creation Labs

A well-designed pilot is the fulcrum of a successful AI adoption program. It is the point at which the abstract becomes concrete, where the promises of the vendor deck are tested against the reality of actual workflows, and where the organization learns what it needs to know to make the program succeed at scale. Misdesigned pilots are treated as rollout previews rather than genuine experiments, run in conditions that do not reflect deployment reality, evaluated on metrics that measure the wrong things, and produce false confidence and costly surprises when scale begins.

The governance of pilots requires the same rigor applied to any material business experiment: defined hypotheses, pre-registered success criteria, controlled conditions, honest reporting of results, including failures, and a decision framework that specifies what results warrant scaling, what warrant iteration, and what warrant stopping. Organizations that run pilots as political exercises designed to produce a predetermined positive result that justifies an already-made investment decision are not learning from their pilots. They are spending money to generate evidence they will cite without examining it.

Designing Pilots as Experiments

Every pilot should begin with at least one falsifiable hypothesis, a statement of what you expect to observe if the system is working as intended, expressed with enough specificity that you could, in principle, observe the opposite. A hypothesis like 'the AI triage system will improve adjuster performance' is not falsifiable in practice; any result can be interpreted as confirming it. A hypothesis like 'adjusters who use AI triage recommendations as first-pass input will close claims with fewer supplemental information requests within 45 days, compared to adjusters using the existing process' is testable, has a clear outcome metric, and a defined comparison group.

The pilot design should explicitly address what 'good enough' means before the pilot begins. This is the success criterion: the minimum result that would justify proceeding to scale. Setting this threshold in advance prevents the common outcome in which initially unimpressive results are declared a success because leadership is committed to the investment. It also prevents the opposite error: abandoning a promising program because early results, which are naturally noisier and smaller than mature program results, fail to meet an unrealistically high standard.

Pilot governance should include a clear failure protocol: if specified thresholds are not met within the defined window, what happens? Options include extending the pilot with additional support, redesigning the underperforming program element, narrowing the scope to the use cases or populations where performance was strongest, or recommending against scaling. Having a documented failure protocol that leadership has endorsed before the pilot begins removes the political pressure to declare success regardless of results. It also signals to pilot participants that the organization is serious about honest learning, which increases the quality of the feedback they provide.

Frontline Co-Creation

Co-creation labs are structured engagements that bring frontline workers into designing the AI experience they will use. They differ from user testing in that participants are not merely evaluating a finished design; they are generating the design. They differ from advisory panels in that their input directly shapes product decisions rather than informing them at a remove. The co-creation model is premised on the insight that the people closest to the work possess knowledge about workflow constraints, cognitive load, professional norms, and decision context that no product team, vendor, or executive can replicate from the outside.

A co-creation lab for an AI adoption program typically runs for four to six weeks. It involves a small cohort of eight to twelve frontline participants working with two to three members of the product team on two to three defined design challenges. The challenges should be real and consequential, the parts of the AI experience that most directly affect user behavior and judgment. For a claims adjuster workflow, the design challenges might include how the triage recommendation should be presented to maximize useful engagement without creating undue anchoring bias; what explanation of the model's reasoning is sufficient to support informed override decisions; and what the audit trail interface needs to look like to be useful in real claims proceedings.

The governance of co-creation labs requires explicit agreements about decision rights. Participants need to understand which of their inputs will be acted upon and which will be considered but potentially overridden by technical constraints, security requirements, or business priorities. Promising participants that they control outcomes they do not control will produce resentment when reality diverges from that promise. Promising participants that their input will be genuinely considered and transparently reported, with explanations for any design choices that diverge from lab recommendations, is both honest and sufficient to secure genuine engagement.

Diagram 5.3 – Co-Creation Lab Workflow

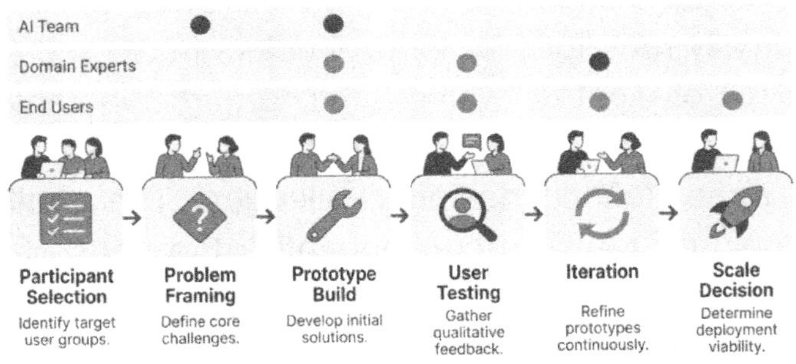

Co-creation labs follow six stages from participant selection through scaled deployment decisions.

The Pilot Checklist

Before launching any pilot, the following conditions should be verified. The pilot hypothesis is written and falsifiable. Success criteria and failure thresholds are defined and endorsed by senior leadership. The comparison group (control arm) is defined and has agreed to maintain current processes during the pilot window. Data collection infrastructure is operational and validated before day one. Participant communication has been delivered, including honest acknowledgment of what the pilot will and will not determine. Escalation paths are defined for participant concerns about data privacy, workflow disruption, or system behavior. A timeline with specific decision points is documented and shared with all stakeholders.

During the pilot, weekly pulse checks should be conducted with a representative sample of participants to capture emerging concerns before they calcify into resistance. These are not satisfaction surveys; they are brief structured conversations with team leaders that surface specific blockers, unexpected workflow interactions, and emerging cultural dynamics. The information captured in weekly pulse checks is often more actionable than the quantitative metrics collected by the telemetry infrastructure, because it contains the qualitative context that makes the quantitative signals interpretable.

At the conclusion of the pilot, results should be reported fully and honestly, including findings that are unflattering or inconclusive, in a structured read-out that addresses each pre-registered hypothesis. The read-out audience should include the frontline participants, not only leadership. Participants who invested their time and credibility in the pilot deserve to know what it found, and transparently communicating the results reinforces the message that the organization is treating the program as a genuine experiment. Selective reporting, sharing positive results with leadership and withholding them from participants or vice versa, destroys the trust infrastructure that co-creation depends on.

5.6 Measurement and Course Correction

AI adoption measurement is not a one-time activity that accompanies a rollout. It is a continuous operational discipline that runs for the life of the program. Organizations that measure intensively during the pilot phase and then revert to passive monitoring after full deployment will miss the second-wave adoption challenges that consistently emerge at the 6-12 month mark: early adopters who plateau, mainstream users who develop workarounds, model drift that degrades the AI's performance without triggering any formal alert, and organizational changes (leadership transitions, workforce turnover) that erode the contextual knowledge that successful adoption was built on.

Sentiment Tracking and Adoption Funnels

Sentiment analytics for AI adoption programs serve a different function than traditional employee engagement surveys. Rather than measuring overall satisfaction with the work environment, they track the specific emotional and cognitive dynamics that predict adoption behavior: perceived relevance of the AI system to daily work, confidence in the accuracy of AI outputs, psychological safety to question or override recommendations, and sense of agency over how the tool is used. These dimensions can be tracked efficiently with a monthly pulse survey of 8-10 questions, supplemented by quarterly qualitative interviews with a stratified sample of users across adoption segments.

Adoption funnel analytics applies the product management concept of conversion funnel analysis to workforce behavior. The funnel has five stages: awareness (does the user know the tool exists and understand its purpose), activation (has the user completed onboarding and made first productive use), engagement (is the user using the tool regularly and in its intended workflow context), proficiency (is the user demonstrating the skill indicators associated with effective use), and advocacy (is the user actively sharing knowledge and recommendations with peers). Tracking the population at each funnel stage and the movement between stages over time gives a much more granular picture of program health than a single activation rate metric.

Usage cohort analysis adds a temporal dimension to funnel analytics. Rather than looking at the user population as a cross-section, cohort analysis tracks groups of users who activated within the same time window as they progress through the funnel. This reveals patterns that cross-sectional analysis obscures: for example, cohorts activated during periods of particularly intensive coaching support may show faster proficiency progression than cohorts activated during the full-scale rollout with reduced coaching attention. This finding has direct implications for decisions about the allocation of coaching resources.

Diagram 5.4 – Adoption Funnel Dashboard

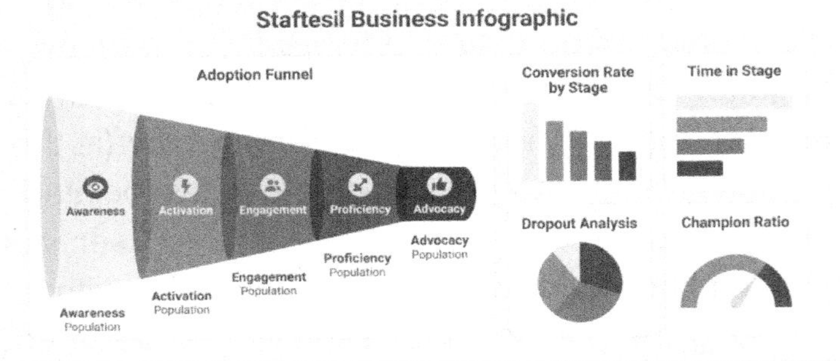

Track five adoption funnel stages to identify where users stall and intervene early.

Rapid Iteration Cycles

Change programs for AI adoption need to be designed for iteration, not just monitoring. There is an important distinction between a program that measures progress and one that acts on those measurements. The former produces dashboards. The latter produces improvement. Building the organizational reflexes for rapid course correction- the ability to identify a program element that is not working, redesign it, test the redesign, and deploy the improved version within a defined timeframe is a core operational competency for change leaders.

A practical, rapid iteration cadence for an AI adoption program runs on three rhythms. Weekly: program operations review, drawing on pulse check data and adoption funnel metrics to identify any emerging issues that require immediate attention. The output is a short action list with owners and due dates. Monthly: program design review, drawing on cohort analysis, sentiment trends, and qualitative user feedback to assess whether program elements are performing as designed. The output is a set of design change recommendations, each with a hypothesis about the mechanism of impact and a measurement plan. Quarterly: strategic review, drawing on outcome metrics and organizational metrics to assess overall program progress against the North Star. The output is a recommendation to senior leadership on program continuation, scale-up, or reorientation.

Each iteration cycle should be distinguished between signal and noise. Program metrics will fluctuate for reasons unrelated to program design: seasonal workflow patterns, external market events, leadership transitions, model updates, and organizational changes all affect adoption metrics independently of the change program. Experienced change managers build this contextual awareness into their interpretation of weekly and monthly data, avoiding the twin errors of overreacting to normal variation and underreacting to genuine trends that are initially small but directionally consistent.

Champion Programs and Success Stories

Champion programs formalize the role of the early adopters and co-creation lab alums who have been acting informally as advocates and coaches throughout the program. A formal champion program provides these individuals with resources, recognition, and organizational legitimacy, and, in return, it provides the program with a scalable network of trusted peer influencers whose reach extends far beyond what any central change management team can achieve.

The design of an effective champion program starts with clarity about what champions are expected to do. The three core activities are: peer coaching (one-on-one and small-group support for colleagues navigating learning challenges), feedback escalation (channeling field-level concerns and product improvement ideas back to the program team through a defined process), and success story documentation (capturing specific examples of workflow improvement, time savings, or quality gains that can be shared with peer audiences). Champions who are asked to do anything beyond these three activities- marketing the program, enforcing compliance, and managing upward- are set up to fail.

Success stories deserve particular attention as a change management tool. The most powerful evidence for the mainstream majority is a specific, verifiable account of a colleague they respect using the AI system to achieve a concrete outcome they care about, not a vendor case study, not an executive testimonial, but a peer narrative. Producing these stories requires investment: identifying them through feedback channels and manager reports, verifying their accuracy, working with the champion to shape them into a compelling format, and distributing them through the channels where the target audience will encounter them. Done well, a library of 15-20 peer success stories is worth more to adoption momentum than any amount of top-down communication.

Diagram 5.5 – Champion Network Architecture

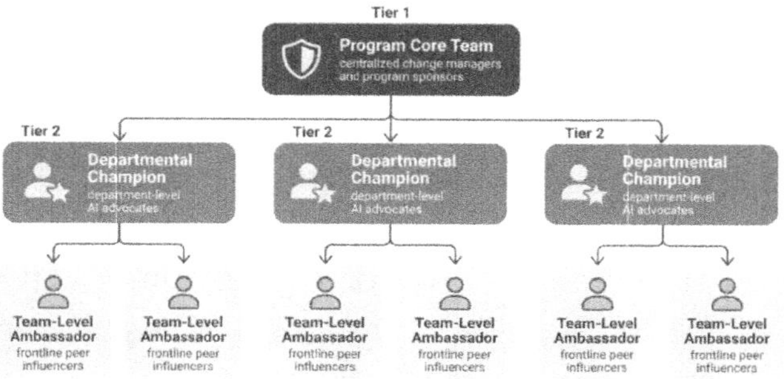

A three-tier champion network extends AI adoption reach from program core to frontline teams.

5.7 Manager's Checklist

Use this checklist to assess your AI change program's design maturity and identify gaps before they become adoption failures:

- North Star definition: Is there a written, falsifiable North Star statement for the program? Has it been tested for resonance with frontline audiences, not just leadership?
- Success metrics: Are success metrics defined at all four layers (adoption, proficiency, outcome, organizational) and pre-registered before the program begins?
- Stakeholder map: Has a stakeholder mapping exercise been completed, producing a segmented picture of early adopters,

mainstream majority, and active resistors by business unit?

- Communication plan: Is there a differentiated communication plan by stakeholder segment, with timing aligned to program milestones and channels matched to audience credibility preferences?
- Middle manager support: Is there a specific communication and support track for middle managers, including clarity on their implementation responsibilities and the resources available to them?
- Pilot hypothesis: Is there at least one falsifiable hypothesis for each pilot, with pre-defined success and failure thresholds endorsed by senior leadership?
- Co-creation participation: Have active skeptics been specifically included in co-creation processes? Are decision rights for co-creation lab outputs clearly defined and communicated?
- Pilot checklist: Have all pre-launch conditions (data infrastructure, escalation paths, participant communication, comparison group agreement) been verified before pilot day one?
- Measurement cadence: Is there a defined weekly, monthly, quarterly measurement cadence with output formats and escalation criteria at each level?
- Champion program: Is there a formal champion program with clearly defined activities (peer

coaching, feedback escalation, success story documentation) and a recognition structure?

5.8 Key Takeaway

AI adoption programs that are led as technology deployments will consistently underperform relative to their potential. Programs that are led as change journeys with rigorous attention to the human dynamics of professional identity, trust, and social influence consistently outperform. The Director of Claims Operations, who turned skeptics into champions, did not do it by explaining the technology better or by mandating its use more forcefully. She did it by creating a structure in which the skeptics' expertise was genuinely valued, their concerns were structurally addressed, and their investment in the program's success was rewarded with ownership and recognition. That structure- the North Star, the stakeholder map, the co-creation lab, and the champion network- is not a soft complement to the technical work. It is the mechanism by which the technical investment produces operational returns. Managers who understand this lead programs that deliver on their promise rather than producing compliance without commitment, which characterizes most AI rollouts.

6 Talent Strategy That Balances Hiring and Reskilling

6.1 Scenario Scaling Fast Without Breaking the Budget

The CIO of a mid-market commercial insurer with roughly 2,800 employees faced a talent problem that no staffing agency could solve. His board had approved an eighteen-month AI transformation roadmap that would touch underwriting, claims, actuarial analysis, and customer service. The assumption embedded in the original budget was that he could staff the effort primarily through external hiring: eight to ten ML engineers, a handful of data scientists, and two or three AI product managers. When he went to market, the reality hit hard. The ML engineers and their technical leads wanted compensation packages that blew past his budget ceiling. Data scientists with insurance domain expertise were rarer than anticipated. And the AI product managers who understood both enterprise product development and modern AI systems were almost nonexistent in his local market. He could have competed for the same talent that every bank, tech firm, and federal contractor was chasing. Instead, he made a different bet. He hired four senior specialists:

two ML engineers with deep model development experience, one AI architect, and one responsible AI program manager. He built a reskilling program that would turn 65 existing employees with strong domain expertise and demonstrated analytical aptitude into capable AI practitioners within 12 months. The results surprised even his own estimates. Nine months in, his internally developed practitioners were outperforming several external hires on the domain-specific tasks that accounted for 80 percent of the program's actual workload. The external specialists provided the foundational technical capability. The reskilled domain experts provided the operational intelligence that made the models useful. Neither group alone would have been sufficient. Together, they were faster and cheaper than a fully external hiring strategy would have been, and they produced an organization that was learning to build AI capability, not just use it.

6.2 Why Talent Strategy Cannot Be an Afterthought

Every AI capability roadmap has a talent dependency that is either explicitly planned or discovered at high cost mid-execution. The roadmap specifies what the organization will build or deploy. The talent strategy specifies who will build and operate it, how they will be sourced, how they will be developed, and how the organization will retain them against competitive pressure. When the talent strategy is not designed alongside the technology roadmap, the most common outcome is a capability gap that becomes visible exactly when the pressure to deliver is highest, after the investment has been made but before the value has been realized.

The talent challenge for enterprise AI programs is fundamentally different from previous technology talent challenges for two structural reasons. First, the supply of practitioners with the specific combination of technical AI capability and enterprise domain knowledge is genuinely limited. The external labor market for ML engineers and AI architects has been tightly contested since 2021, and the practitioners most sought after are those who can translate business problems into model behaviors, who understand both technical constraints and operational context, and are among the rarest profiles in the workforce. Second, the window for competitive differentiation through talent is measured in months, not years. Organizations that cannot staff AI programs within their planning timeline find themselves watching competitors who can.

These structural realities make a balanced talent strategy one that combines targeted external hiring for frontier capabilities with systematic reskilling of existing employees for scale roles, not merely a cost-saving option, but often the only viable path to program velocity. Pure external hiring strategies are budget- and time-constrained and retention-challenged. Pure reskilling strategies cannot fill genuine technical gaps that require years of specialized development. The design question is not about hiring versus reskilling, but about determining the right blend for each role family and executing both tracks with the discipline required to meet the program's timeline.

The sections that follow provide a framework for making this determination systematically. The framework covers talent segmentation, the hiring playbook for frontier roles, the design of reskilling programs for scale roles, mobility and rotation mechanisms, external partnerships, and the measurement infrastructure needed to manage talent strategy with the same rigor applied to any other program investment.

6.3 Talent Segmentation Frontier Roles and Scale Roles

The foundational move in AI talent strategy is the distinction between frontier roles and scale roles. This distinction is not about seniority or compensation; a scale role can be highly senior and highly compensated. It is about the source of the required capability and the strategic logic for its best acquisition and development.

Frontier roles require capabilities that are in limited supply in the external market and cannot be practically developed through internal reskilling on a program-relevant timeline. They are typically the smallest segment of the talent needed, often 10-20 percent of the AI program's total headcount. Still, they are disproportionately critical because they provide the technical foundations on which the rest of the program depends. Getting the frontier role hiring wrong can delay and sometimes derail well-designed programs.

Scale roles require capabilities that can be developed through a well-designed reskilling program starting from a foundation of domain expertise, analytical aptitude, and organizational knowledge that internal candidates already possess. They represent the majority of the talent needed for most enterprise AI programs. Getting scale role reskilling right determines whether the program builds durable organizational capability or becomes dependent on external talent that can leave at short notice.

Identifying Frontier Roles

Identifying frontier roles requires answering two questions for each role family in the AI program workforce: First, could an internal candidate with strong domain knowledge and demonstrated analytical aptitude develop the required capability within 12-18 months with structured support? Second, is the external supply of qualified candidates sufficient to fill this role within the program's hiring window at sustainable compensation levels? When the answer to both questions is no, you have a frontier role that requires a purpose-built hiring strategy.

Common frontier roles for enterprise AI programs include ML infrastructure engineers (the practitioners who build and maintain the training, inference, and monitoring infrastructure that AI systems run on), AI safety and alignment specialists (who design the responsible-by-design systems that govern model behavior in high-stakes operational contexts), and AI product architects (who bridge model capabilities and product requirements with the depth needed to make consequential design decisions). These roles are in short supply, not because they require genius, but because they require a specific combination of technical depth and practical experience that takes years to develop and is scarce relative to demand.

A useful heuristic for distinguishing frontier from scale roles is the 'context premium': how much of the role's value comes from deep knowledge of your specific operational context, your claims workflow, your underwriting criteria, your patient population, versus universal technical skills that transfer across contexts? Roles with a high context premium are better candidates for reskilling from internal talent, while roles where universal technical skills dominate are better candidates for external hiring. Most AI program roles sit somewhere on the spectrum; the segmentation exercise places them on the right side of the hiring-versus-reskilling threshold.

Defining Scale Roles

Scale roles determine whether the AI program's benefits are realized across the organization rather than confined to a small technical team. They include AI-augmented practitioners (domain experts who use AI tools as a core part of their workflow and are expected to develop increasing sophistication over time), workflow automation designers (practitioners who map current-state processes and redesign them around AI capabilities), model output validators (specialists who assess whether AI recommendations meet quality, accuracy, and fairness thresholds before they inform operational decisions), and AI program managers (practitioners who coordinate delivery across technical and operational workstreams with sufficient AI literacy to manage dependencies and surface risks).

Defining scale roles precisely is important because precision is what makes reskilling programs targetable. A vague target, 'we need people who are good with AI,' cannot be turned into a curriculum, an assessment framework, or a hiring profile. A specific target, 'we need underwriters who can evaluate AI risk scores against case-specific contextual factors and document their reasoning in a format that satisfies audit requirements,' can be developed into a learning program, a competency assessment, and a career pathway.

Scale role definitions should be co-authored by three groups: the AI technical team (who understand what the AI system produces and what capability is needed to use it well), the business operations team (who understand the workflow context and the performance standards that apply), and the people operations team (who understand the development pathways and career implications). Role definitions produced by any one of these groups in isolation will be either technically accurate but operationally impractical, operationally grounded but technically shallow, or organizationally coherent but disconnected from the actual capability requirements.

Diagram 6.1 – Talent Segmentation Framework

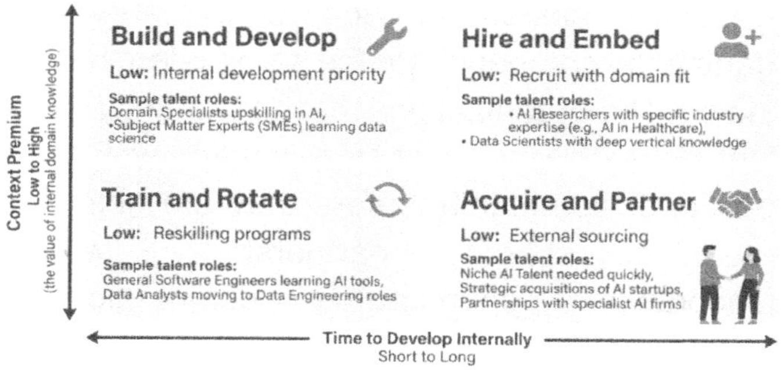

Segment AI talent needs by context value and development time to prioritize build-versus-buy decisions.

6.4 The Hiring Playbook

A hiring playbook for frontier AI roles must operate differently from a standard enterprise recruiting process. Standard enterprise recruiting is optimized for volume, compliance, and consistency. It works well when the talent profile is well-understood, the supply is adequate, and the assessment can rely on established credential signals. For frontier AI roles, none of these conditions hold. The talent profile is evolving, the supply is constrained, and the most relevant credentials, such as open-source contributions, demonstrated project outcomes, and novel problem-solving capability, are poorly captured by traditional résumé screening and interview processes.

Sourcing and Partnerships

Effective sourcing for frontier AI roles requires a proactive pipeline strategy rather than a reactive job-posting strategy. By the time a frontier role requirement is actively posted on job boards, every other organization competing for the same talent has already done so. The organizations that consistently win in tight talent markets have built relationships with the target practitioner community before they are actively hiring through research partnerships, conference sponsorships, open-source contributions, and technical meetup hosting.

University research partnerships provide early access to graduate students and postdoctoral researchers who represent the next cohort of frontier practitioners before they enter the competitive labor market. A partnership with a computer science department that includes co-advised research projects, internship pipelines, and curriculum advisory participation costs far less than the recruiting fees and compensation premiums required to win competitive hires in the open market. It also provides access to faculty advisors' academic networks, a high-signal source of referrals for senior practitioners.

Internal referral networks are consistently the highest-quality source of candidates for frontier roles and are chronically underleveraged by enterprise HR functions that disproportionately focus on external channels. The two or three frontier practitioners already within the organization know the practitioner community, have credibility within it, and can assess candidates with precision that recruiting coordinators cannot match. Building a referral program that provides meaningful incentives and that makes the referral process low-friction for technical employees is a high-ROI investment in sourcing infrastructure.

Compensation and Career Ladders

Compensation for frontier AI roles in enterprise environments presents a genuine structural challenge. The market-clearing price for experienced ML engineers and AI architects at technology companies frequently exceeds the compensation bands used by mid-market and public-sector organizations for their highest individual contributor and senior manager roles. Organizations that try to compete at market rate for all frontier roles will strain or break their compensation architecture. Organizations that refuse to pay market rates will consistently lose competitive candidates to technology firms.

The practical resolution is a two-track approach. The first track applies market-rate compensation selectively to the smallest possible set of genuinely frontier roles, the four or five practitioners whose specific capabilities are irreplaceable and whose market alternatives are most competitive. This requires adjusting compensation bands for a defined set of 'specialist' or 'distinguished contributor' roles that sit outside the standard individual contributor ladder. Most governing boards and HR policy committees will accept this adjustment if the business case is presented clearly: a small number of above-band roles is less costly than the program delay caused by being unable to hire.

The second track offers non-compensation differentiation to candidates for whom mission, organizational culture, and development opportunity are significant factors in their employment decision. Public sector AI programs have found particular success in attracting frontier talent by framing the mission: practitioners who want their work to improve public services, reduce health disparities, or increase government effectiveness, and who sometimes accept below-market compensation in exchange for mission alignment and the technical challenges that large-scale public systems offer. The key is to understand which candidates this applies to and to invest in making the mission credible, with specifics rather than slogans.

Diagram 6.2 – AI Career Ladder Model

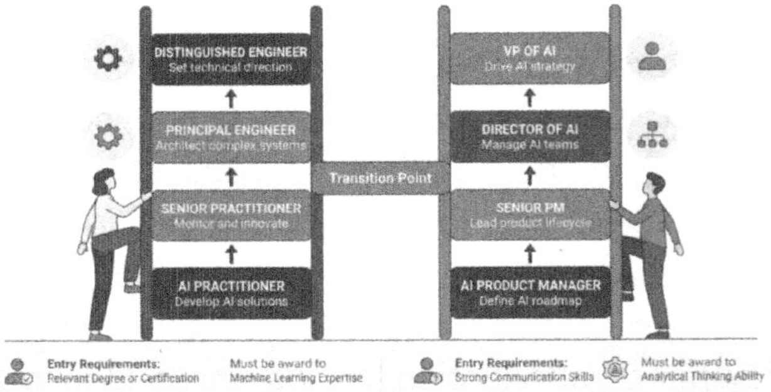

Dual AI career tracks give technical and leadership professionals clear advancement paths

Promotion Criteria for AI Fluency

One of the most powerful signals an organization can send about the importance of AI capability development is incorporating AI fluency into formal promotion criteria. When AI proficiency is a stated requirement for advancement, not an informal preference, but an explicit criterion in the promotion rubric, voluntary participation in reskilling programs accelerates dramatically. This is the organizational behavior equivalent of what micro-credentials do at the individual level: it makes the career consequence of capability development legible and credible.

Promotion criteria for AI fluency should be graded by level and role family. For entry- and mid-level roles, AI fluency criteria might require demonstrated proficiency with AI tools in the relevant workflow context, successful completion of Tier 1 and Tier 2 micro-credentials, and at least one documented example of using AI outputs to improve a work product or decision. For senior and principal roles, criteria might include AI program leadership experience, a demonstrated ability to identify and design new AI use cases, and evidence of contributions to the organization's AI knowledge commons through mentoring, documentation, and peer review.

Introducing AI fluency into promotion criteria requires careful implementation to avoid creating a two-tier workforce dynamic in which employees who had extended leave, part-time arrangements, or role-specific constraints during the reskilling program window are disadvantaged. Grace periods, alternative demonstration pathways, and reasonable accommodation provisions should be built into the criteria from the outset, rather than retrofitted after a grievance or legal challenge forces the issue.

6.5 Mobility and Rotation Programs

Talent strategy for AI programs is not solely about filling roles. It is about the continuous movement of capability across the organization, ensuring that AI expertise developed in one part of the business flows to other parts, that practitioners' skills are sharpened by exposure to multiple contexts, and that the organization builds a diverse bench of AI-capable leaders rather than concentrating expertise in a single team that becomes a bottleneck and a retention risk.

Mobility programs that establish mechanisms to rotate practitioners across business units, functions, and project types are the organizational design equivalent of portfolio diversification in investment management. They reduce the risk of concentration, accelerate knowledge transfer, and build the cross-functional relationships that complex AI programs require. Organizations that lack mobility programs tend to develop AI talent silos: pockets of excellence surrounded by deserts of capability, connected by informal networks that depend on specific individuals staying in specific roles.

Secondments and Stretch Assignments

A secondment is a temporary assignment in which a practitioner moves from their home team to a different business unit or functions for a defined period, typically three to twelve months, before returning. For AI talent development, secondments serve several functions simultaneously. They transfer capability: the seconded practitioner brings their AI expertise to a part of the organization that has less of it. They develop the practitioner: exposure to a different operational context and stakeholder set builds the domain breadth and organizational intelligence that distinguishes senior practitioners from technically narrow specialists. And they build relationships: practitioners who have worked across multiple business units serve as the connective tissue of the AI organization, enabling the informal information flows that formal governance structures cannot fully replace.

Secondment design requires attention to the sending team's incentive structures. A team lead who loses a valued practitioner to a six-month secondment absorbs a real productivity cost. If that cost is fully borne by the sending team, with no organizational compensation, no backfill, no recognition, and no priority access to future secondment talent from other units, the team leader will find ways to block the secondment or send their least valuable practitioners rather than their most capable. Secondment programs that work have explicit funding mechanisms (the program, not the sending team, covers backfill costs), clear priority queues for receiving future rotation talent, and recognition structures that credit team leads for contributing to organizational capability development.

Stretch assignments are a lighter-weight mobility mechanism: a practitioner takes on a project or responsibility that requires capabilities beyond their current role profile, typically while remaining in their home team. For AI capability development, stretch assignments might include leading a use-case discovery workshop in an unfamiliar business unit, serving as the AI program lead for a pilot deployment, or co-designing a training module for a micro-credential program. Stretch assignments build capability, signal organizational trust, and create a portfolio of demonstrated accomplishments that supports promotion cases, all at lower organizational cost than a full secondment.

Mobility Metrics

Mobility programs should be managed with the same rigor as other talent investments, which requires a measurement framework that tracks their costs and impact. The core mobility metrics for an AI talent program are: internal mobility rate (what percentage of AI program roles are filled by internal candidates, including secondments and stretches), time to productivity for rotated practitioners (how quickly practitioners become fully productive in a new context, compared to external hires in similar roles), knowledge transfer effectiveness (measured through pre- and post-assessments of AI capability in receiving teams), and retention impact (whether practitioners who have participated in mobility programs show higher retention rates than those who have not).

The retention impact metric deserves particular attention. Mobility consistently ranks among the top factors AI practitioners cite for staying with an employer rather than pursuing external opportunities. Practitioners who see a clear path to varied, challenging assignments within the organization are less likely to conclude that the only way to advance is to leave. Quantifying this retention benefit, even approximately, by tracking the retention differential between mobile and non-mobile practitioners, provides a financial case for mobility program investment that resonates with finance-focused leadership.

Diagram 6.3 – Rotation Program Architecture

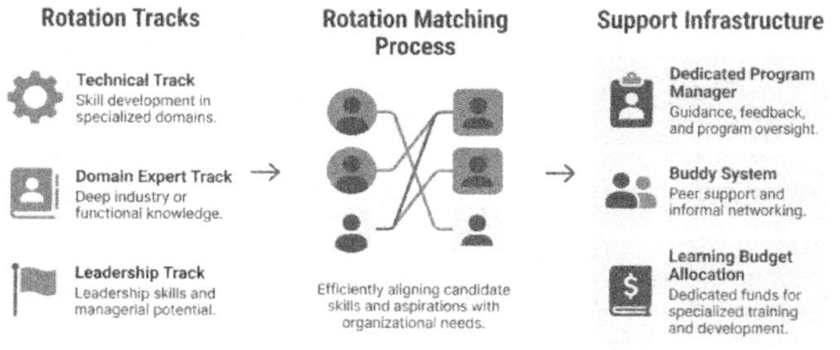

Rotation program architecture combines defined tracks, smart matching, and dedicated support infrastructure.

6.6 External Partnerships and Contingent Talent

No talent strategy for enterprise AI is complete without an external partnership layer. The internal talent pipeline, whether developed through reskilling or built through hiring, cannot meet every capability need on every timeline. External partnerships extend the organization's reach into the talent ecosystem in ways that pure employment strategies cannot. They provide access to specialized expertise that may not warrant a full-time role, connections to emerging research that keep the program's technical approach current, and credentialing infrastructure that supports the micro-credential architecture described in earlier sections.

The governance of external partnerships requires as much discipline as the governance of internal talent programs. Partnerships that are not actively managed, exist as signed agreements without ongoing engagement mechanisms, shared governance, and outcome measurement, will drift toward underutilization or become sources of dependency that the organization did not intend to create. Responsible-by-design partnership governance includes defined deliverables, regular touchpoints, clear intellectual property agreements, and exit provisions that ensure the organization is not left stranded if a partner relationship ends.

University and Bootcamp Partnerships

University partnerships for AI talent development serve three distinct functions: early-stage talent pipeline (access to graduate students and recent graduates before they enter the competitive external market), research collaboration (joint work on applied AI problems that advances organizational capability and builds relationships with academic practitioners), and curriculum development (co-designing the content of internal learning programs with academic partners who bring pedagogical depth and theoretical rigor).

Effective university partnerships require investment on both sides. The university partner needs resources (funding for research positions, computing infrastructure, and data access under appropriate governance), recognition (co-authorship and publication rights for any joint research, and participation in industry advisory roles that count toward faculty tenure and promotion), and genuine intellectual engagement rather than purely transactional project work. Partnerships designed purely as low-cost recruiting pipelines will be recognized as such by faculty and treated accordingly: the best students and researchers will choose to work with partners who take the collaborative dimension seriously.

Bootcamp partnerships serve a different function than university partnerships: they provide rapid-development pathways for existing employees who need to build specific technical skills on a program-relevant timeline. A six-month intensive bootcamp in data engineering or machine learning fundamentals can bring a strong domain expert to the threshold of a scale role in a timeframe that a self-directed learning program cannot match. Evaluating bootcamp partners requires attention to curriculum relevance (is the program building the skills your scale roles require, or generic skills that will need context-specific supplementation), completion and placement rates (what is the actual track record of participants achieving the intended capability milestone), and employer engagement (does the bootcamp incorporate your organization's specific use cases and tools into the curriculum).

Vendor Selection and SOW Design

Contingent talent contractors, staff augmentation, and managed service teams are necessary components of most enterprise AI program talent mixes, particularly during the initial build phase, when capability needs are highest and internal capacity is still being developed. Managing contingent AI talent responsibly requires more rigorous contracting than standard staff augmentation arrangements, because the work product is higher-stakes and the implications for knowledge transfer are more significant.

The design of a Statement of Work for AI-related contingent talent should address four elements that standard SOW templates typically omit. First, knowledge transfer requirements: the SOW should specify which documentation, training, and handoff activities the vendor is obligated to perform and when, to ensure that internally developed capabilities can sustain the work product after the engagement ends. This is not about replacing the vendor; it is about ensuring that the value created during the engagement does not depart with the vendor's team. Second, data handling and model ownership: the SOW should be explicit about who owns any models, training data, and derived insights produced during the engagement, as well as the vendor's obligations regarding data security, retention, and deletion.

Third, quality and performance standards: unlike many professional services engagements, where deliverables are documents or project plans, AI engagements often produce systems whose quality can only be assessed over time through performance measurement. The SOW should define the applicable performance metrics, the measurement period, and the remediation obligations if performance falls below the threshold. Fourth, responsible-AI compliance: the SOW should incorporate your organization's responsible-AI standards as binding requirements rather than mere guidelines, with defined audit rights to verify compliance.

Centers of Excellence

A Center of Excellence (CoE) is an organizational structure that centralizes frontier capabilities within a shared service model, making them available to multiple business units without requiring each unit to build its own deep technical team. For AI programs, the

CoE model resolves a tension that is otherwise very difficult to manage: frontier AI talent is expensive, in short supply, and most effective when applied to complex problems that require deep focus. If distributed across business units in ones and twos, frontier practitioners spend most of their time on routine support tasks that do not require their capabilities and have little opportunity for the cross-pollination that produces technical creativity. A CoE concentrates them in an environment where they can collaborate, tackle complex problems, and provide structured support to business unit teams through a defined engagement model.

CoE design requires careful attention to the engagement model, the mechanism by which business unit teams access CoE services. The two most common models are the project-based model (business units submit requests for specific projects, and the CoE assigns team capacity to fulfill them) and the embedded liaison model (CoE practitioners spend a defined portion of their time embedded in each business unit, providing ongoing support within a defined scope while maintaining their primary affiliation and professional community in the CoE). The project-based model provides efficient utilization but creates a transactional dynamic that can limit knowledge transfer. The liaison model creates richer relationships but requires practitioners who can operate effectively across multiple organizational contexts.

CoE governance must address the tension between organizational responsiveness and technical focus. Business units want CoE capacity on demand. CoE practitioners need protected time to develop the deep work model, conduct safety evaluations, and design the architecture, which are their primary value contributions. The resolution is a defined capacity-allocation framework: a specified percentage of CoE capacity is reserved for deep work and is not interruptible by business-unit requests, and a specified percentage is allocated to business-unit engagements through a prioritized queue. This framework needs to be endorsed by senior leadership and enforced by the CoE director, who has sufficient organizational authority to push back on demands that exceed allocated capacity.

Diagram 6.4 – Partnership Ecosystem Map

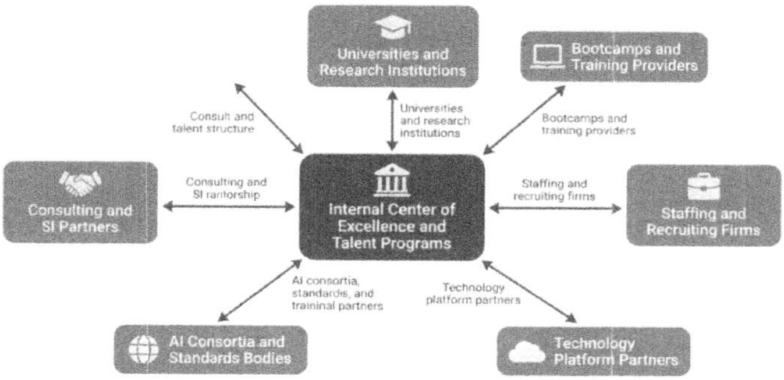

Map your AI talent partnership ecosystem across six external partner categories

6.7 Manager's Checklist

Use this checklist to assess the completeness and rigor of your AI talent strategy:

- Talent segmentation: Have you completed a systematic segmentation exercise that classifies each AI program role family as frontier, scale, or hybrid, with a documented rationale for each classification?
- Frontier role sourcing: Is there an active pipeline strategy for frontier roles, including university partnerships, internal referral programs, and technical community presence, rather than a reactive job-posting approach?
- Compensation architecture: Is there an approved compensation mechanism for frontier roles that is competitive with the external market, without requiring above-band compensation for scale roles?
- Career ladder: Is there a published, differentiated career ladder for AI roles covering both the technical track and the product/program track, with defined competencies and micro-credential alignment at each level?
- Promotion criteria: Have AI fluency criteria been incorporated into the promotion rubric for all role families where AI proficiency is material to performance? Are grace periods and alternative demonstration pathways defined?

- Reskilling program: Is there a structured reskilling program for scale roles with defined cohorts, assessed outcomes, and career consequences for completion? Is it targeted at the specific capability gaps identified in the segmentation exercise?
- Rotation program: Is there an active mobility program with defined rotation tracks, backfill funding, and governance for assignment matching? Are mobility metrics tracked and reviewed quarterly?
- University partnerships: Is there at least one active university partnership with defined deliverables in all three function areas (pipeline, research, curriculum)?
- Contingent talent governance: Do active SOWs for AI-related contingent talent include knowledge transfer requirements, model and data ownership provisions, performance standards, and responsible-AI compliance obligations?
- CoE design: If a Center of Excellence exists or is planned, is there a defined capacity allocation framework with protected deep-work time and a prioritized business unit engagement queue endorsed by senior leadership?

Diagram 6.5 – Talent Strategy Measurement Dashboard

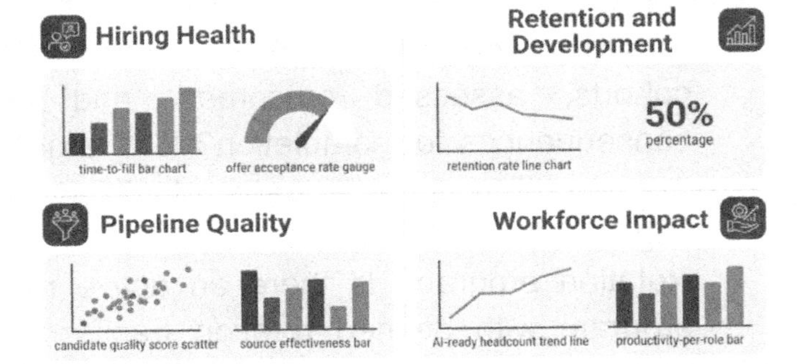

Track AI talent strategy health across hiring, retention, pipeline quality, and workforce impact.

6.8 Key Takeaway

The organizations that scale AI programs successfully are not the ones that win every talent competition in the external labor market. They are the ones who build a talent strategy that is mission-aligned, structurally sound, and continuously managed. This means being disciplined about which roles genuinely require external frontier talent and which roles are better filled by reskilling domain experts who already possess the organizational knowledge that makes AI systems useful in context. It means designing career ladders and promotion criteria that make AI capability development a rational personal investment for every employee, not just the technically inclined minority. It means building mobility programs that spread capability across the organization and reduce the risk of concentration that comes from relying on a small number of irreplaceable practitioners. And it means governing external partnerships with universities, bootcamps, vendors, and contingent staffing with the rigor the stakes demand. The CIO who blended four specialist hires with a 65-person reskilling program did not stumble onto a talent strategy by accident. He made a deliberate bet based on a systematic analysis of where external supply was constrained, where internal potential was underutilized, and what combination would produce the fastest path to operational capability at sustainable cost. That analytical discipline, applied consistently across every talent decision, is what separates AI programs that build durable organizational advantage from those that produce impressive launch announcements followed by quiet capability gaps.

7 Platform Enablement That Keeps Product Teams Accountable

7.1 Scenario: The Platform That Cut Integration Time in Half

The VP of Engineering at a regional health system had watched the same problem repeat itself for three years. Every time a product team wanted to add an AI-assisted feature, a prior authorization recommender, a discharge risk scorer, or a claims review aid, they rebuilt the same scaffolding from scratch. They pulled their own data pipelines. They set up their own model-serving containers. They wrote their own monitoring dashboards. Each team operated like an expedition packing its own oxygen, even though the base camp had tanks to spare.

The integration timelines averaged fourteen weeks from approved concept to production deployment. That lag was not an engineering problem in the ordinary sense. It was a platform problem. The health system had no shared feature store, no canonical model registry, and no standardized monitoring hooks. Product teams owned their own pipelines end-to-end, which felt empowering until it created a system of fragile, incompatible, unauditable AI components scattered across the organization.

The VP chartered a platform engineering team with a clear mandate: reduce time-to-production for new AI-assisted features without removing product team autonomy. Eighteen months later, the average integration time had dropped from 14 weeks to 6 weeks. Duplicate feature computation had been eliminated across four product areas. Monitoring coverage went from ad hoc to universal. The platform team had not taken control of AI decisions; they had made the right decisions easier to implement. That is the job.

7.2 Why Platforms Must Balance Efficiency and Autonomy

A platform exists to multiply the productivity of teams building on top of it. In AI-enabled organizations, the platform is not optional infrastructure; it is the mechanism through which standards, governance, and reuse become real rather than aspirational. Without a platform, every product team invents its own approach to data access, model deployment, and quality monitoring. Some of those inventions will be excellent. Most will be inconsistent. A few will be dangerous.

The opposing risk is a platform that stifles the experimentation that makes AI valuable. If every model deployment requires a six-week review cycle and three approval committees, product teams will route around the platform. They will spin up shadow infrastructure in personal cloud accounts or use vendor-managed APIs that never touch central governance. The platform becomes irrelevant to the work that matters most, and the organization loses the visibility it needs to manage AI risk.

The correct design posture is constrained autonomy: product teams can move quickly within well-defined boundaries, and those boundaries are enforced by the platform rather than by committee. This distinction is important. When governance is enforced by structure, by the fact that the platform only accepts model artifacts registered in a certain format, or that monitoring is automatically attached to every deployment, compliance becomes the path of least resistance rather than an extra burden. Responsible by design means the platform architecture itself carries governance load.

This chapter provides a framework for building and operating such a platform. The target audience is the manager who oversees the platform team, the product leader who depends on it, or the CIO trying to understand how to fund and govern it. The decisions here are not primarily technical. They are organizational: what does the platform own, what do product teams own, and how do you resolve disputes when those boundaries are tested?

Platform maturity is not a one-time achievement. It is a continuous negotiation between standardization and flexibility, between the needs of teams building today and those of a governance regime that must endure over the years. Managers who treat the platform as a project with a completion date will consistently be surprised when it needs to evolve. Managers who treat it as an ongoing product with its own roadmap, users, and success metrics will build something durable.

7.3 Defining Platform Primitives

Platform primitives are the foundational capabilities that every AI product team needs but that no individual team should have to build from scratch. In AI-enabled organizations, three primitives carry most of the weight: the feature store, the model registry, and monitoring hooks. Getting these three right lays a foundation for consistently, auditably, and safely building any number of AI-assisted products. Getting them wrong or skipping them means that every team's AI capability is essentially a one-off, disconnected from the organization's ability to oversee, reuse, or improve it.

The sequence in which organizations adopt these primitives matters. Most teams can absorb a feature store before they can use a model registry effectively, and they cannot meaningfully use monitoring hooks until they have models in production worth monitoring. A realistic platform roadmap sequences these primitives in that order, while building the organizational practices that make each one stick.

Diagram 7.1 – Platform Primitive Architecture

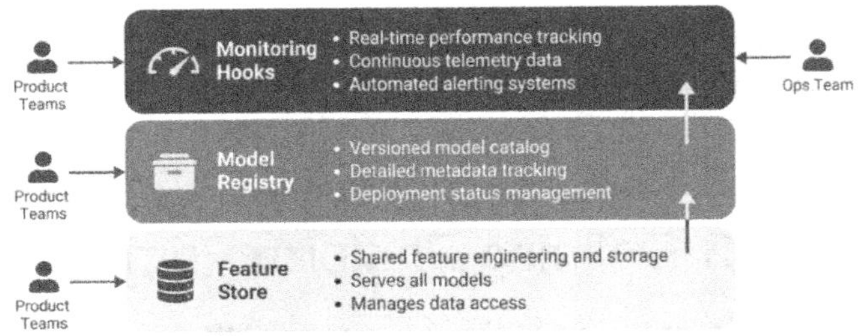

Three platform primitives - feature store, model registry, and monitoring - serve all AI products.

Feature Store

A feature store is a centralized repository for the computed signals that models use as inputs. Without a feature store, teams compute the same features in parallel: one team computes the thirty-day rolling claims volume for their fraud model, another computes it for their utilization predictor, and a third computes a slightly different version for their authorization recommender. These computations diverge over time as each team adds its own caching logic, data freshness rules, and edge-case handling. The result is three models trained on subtly different definitions of the same concept, with no mechanism to detect or correct the drift.

A feature store solves this by separating feature computation from feature consumption. The platform team, or a designated data engineering function, is responsible for computing canonical features once and making them available to any model that needs them. Product teams register their feature requests with the store and consume precomputed values during training and inference. This separation reduces duplication, improves consistency, and enables tracking exactly which data each model was trained on, a critical requirement for auditability in regulated industries.

The feature store also enables training-serving consistency, which is one of the most common and most underestimated failure modes in production AI. When the feature computation logic used to train a model differs even slightly from the logic used at inference time, the model receives inputs it was never trained to handle. Predictions degrade silently. A shared feature store enforces consistency because the same computation artifact is used in both contexts.

For managers, the key decisions around a feature store are organizational rather than technical. Who owns the canonical definitions of shared features? What is the process for a product team to propose a new feature or dispute an existing definition? How are features versioned when the underlying data schema changes? These questions require governance answers before they require engineering answers. A feature store without a feature governance process is just a database with an API.

Adoption metrics for a feature store should track the percentage of models in production that consume features from the store rather than compute them themselves. This ratio is a direct measure of platform leverage. Organizations that achieve more than seventy percent coverage can begin to treat their feature store as a genuine source of organizational knowledge about their data, not just a technical convenience, but a record of what the organization has learned to measure.

Model Registry

A model registry is a versioned catalog of trained model artifacts, along with the metadata required to understand, reproduce, and govern each one. At a minimum, a registry entry should capture the model version, training data lineage, evaluation metrics, the team that owns it, the use case it supports, and its current deployment status. In regulated contexts, it should also capture the approval record for any model deployed to production.

The model registry is what makes governance operational. Without it, questions like 'which version of the authorization model is currently running in production?' or 'when was the fraud model last retrained and what data was it trained on?' require a detective operation across engineering wikis, Slack threads, and personal notebooks. With a registry, those answers are available in seconds. This is not a convenience; it is the operational foundation for incident response, audit, and regulatory examination.

The registry also enables promotion workflows. Rather than allowing any engineer to deploy any model artifact directly to production, a registry-based workflow requires that a model be registered and reviewed before it can be promoted through staging environments. The review can be lightweight, a peer engineer checking that evaluation metrics meet the defined threshold, or it can involve formal approval from a risk or compliance function for higher-stakes use cases. The platform enforces the workflow; the organization defines what the workflow requires.

For managers overseeing platform adoption, the model registry also serves as a portfolio management tool. It provides visibility into how many models are in production, how old they are, which ones are actively maintained, and which ones have drifted outside acceptable performance bounds. Without that visibility, an AI portfolio is effectively invisible to the managers responsible for it. A registry-based approach makes the portfolio legible, which is a prerequisite for managing it with operational clarity.

One common adoption challenge is getting product teams to treat the registry as the authoritative source of truth rather than as an optional reporting requirement. The most effective mechanism is to make registry registration a hard prerequisite for platform-provided services. An unregistered model cannot receive monitoring. An unregistered model cannot be deployed to the platform's serving infrastructure. When registration is the gate, teams register. When registration is optional, it is inconsistent.

Monitoring Hooks

Monitoring hooks are the standardized interfaces through which the platform collects signals about model behavior in production. They are 'hooks' rather than 'monitors' because the platform provides the collection mechanism while product teams define what matters in their context. The hook captures prediction volume, latency, and input and output distribution statistics. The product team configures the alert thresholds and dashboard views that make those signals meaningful for their use case.

This division of responsibility matters for adoption. If the platform team tries to define what constitutes a problem for every use case, alert thresholds, drift definitions, and acceptable error rates, they will be wrong in many cases. The fraud model and the clinical risk model have completely different tolerance profiles. The platform team cannot know those profiles from the outside. But if monitoring is left entirely to product teams, some will do it well, and most will do it minimally. Standardized hooks with configurable thresholds give product teams meaningful control without allowing them to opt out of monitoring entirely.

At the organizational level, monitoring hooks create the shared signal that enables the platform team to maintain a system-wide view of AI health. When all models emit the same classes of metrics to the same observability infrastructure, it becomes possible to build dashboards that show the health of the entire AI portfolio, not just individual models. That portfolio view is the operational instrument that managers need to fulfill their oversight responsibilities without requiring detailed technical involvement in each product.

The monitoring hook capability should also extend to data drift detection. When the distribution of model inputs changes due to a system change upstream, a seasonal shift in behavior, or the onboarding of a new data source, the model's performance may degrade without an obvious error signal. Data drift detection identifies these shifts before they translate into outcome degradation, giving product teams time to investigate and retrain before the problem affects users. This is a classic example of the platform's workflow-level impact that most individual product teams could not build themselves.

7.4 Setting Boundaries Product Autonomy with Guardrails

The most contested governance question in AI platform design is not technical; it is jurisdictional. What decisions belong to the platform team, and what decisions belong to the product team? When those boundaries are ambiguous, every significant decision becomes a negotiation. When they are clear, most decisions can be made without escalation, and the escalation process for genuinely contested cases is more productive because the starting positions are well understood.

The boundary framework presented here is designed to maximize product team velocity while preserving the platform team's ability to maintain safety, consistency, and auditability across the organization. It is not a fixed rule; it should be revisited as the platform matures and as the organization's AI capabilities develop. But it provides a stable starting point that most enterprise organizations can adopt without major reorganization.

Diagram 7.2 – Product-Platform Boundary Map

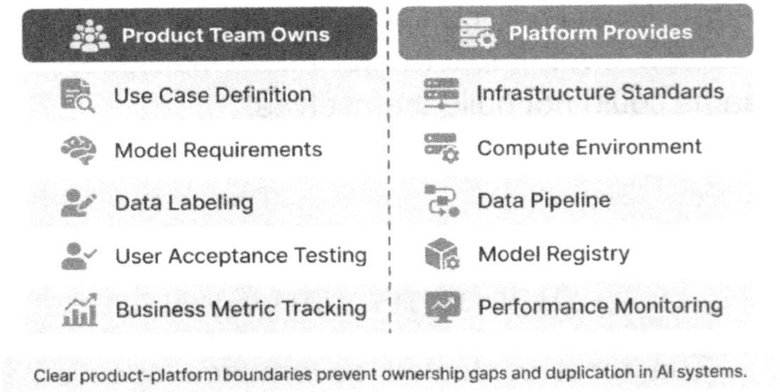

What Product Teams Own

Product teams own the use case and the outcome. They define the problem the model is solving, the population it affects, the success metrics that matter to the business, and the user experience through which model outputs are surfaced. They own the decision to build an AI-assisted feature in the first place, the evaluation criteria that determine whether a model is good enough to deploy, and communication to users and stakeholders when something goes wrong.

Product teams also own the configuration of platform services for their use case. They configure the alert thresholds on monitoring hooks. They specify the data freshness requirements for their feature consumption. They define the rollback criteria that are triggered when a production model degrades. The platform provides the mechanism; the product team provides the judgment about what the mechanism should do in their context.

Product team ownership extends to the business case for AI investment. The platform team should not be in the business of evaluating whether a given AI feature is worth building; that is the product team's domain, validated by their product leadership chain. The platform team's job is to make sure that whatever the product team builds can be built safely and consistently, not to filter which ideas are worth pursuing.

One nuance that experienced platform managers learn to navigate is that product teams often want to own the infrastructure decisions the platform needs to standardize on. A product team that prefers a particular model-serving framework, feature computation library, or monitoring tool will resist platform-wide standardization of those choices. The resolution is to distinguish between essential standardization, where divergence creates governance risk or integration cost, and incidental standardization, where the platform is just imposing a preference. Product teams should win incidental debates. The platform should hold on to the essential ones.

What the Platform Provides

The platform provides the shared infrastructure that makes every product team's AI work safer, faster, and more consistent. This includes the feature store, the model registry, the serving infrastructure, the monitoring and collection layer, the governance workflow tooling, and the cost-tracking system. These are not optional services the platform offers; they are the platform's core deliverables, and every AI-assisted product deployed in the organization should use them.

The platform also provides the guardrails that enforce organizational standards without requiring manual review of every decision. Cost quotas limit spending without requiring approval for every experiment. Approval workflows for high-stakes use cases ensure that governance review occurs without slowing standard deployments. Data access controls ensure that sensitive data is available only to models that have been appropriately reviewed, not through individual data access requests, but through the platform's access control model.

Support is a platform product, not an afterthought. The platform team should maintain a tiered support model: self-service documentation and runbooks for common tasks, async support with defined SLAs for integration questions, and an escalation path for production incidents involving platform infrastructure. Product teams that cannot get platform support when they need it will solve their problems by working around the platform, which defeats its purpose. Support SLA is a platform accountability mechanism.

The platform team should also maintain a visible roadmap for product teams. When product teams understand what platform capabilities are coming and when, they can plan their own roadmaps accordingly. When the platform roadmap is opaque, product teams make conservative assumptions, either waiting indefinitely for capabilities that would simplify their work or building their own versions of capabilities that the platform is about to ship. Both outcomes are wasteful. Platform roadmap transparency is a multiplier on product team efficiency.

7.5 Self-Service Enablement

Self-service is not a feature of a mature platform; it is the goal of platform design. A platform that requires product teams to open tickets and wait for platform engineers to fulfill requests is not a platform; it is a service bureau. Service bureaus create bottlenecks, generate resentment, and do not scale. A genuine platform enables a product team to go from an approved concept to a staged deployment without requiring any manual assistance from the platform team, provided the concept falls within established parameters.

Reaching genuine self-service capability requires investment in three areas: templates that package the platform's best practices into reusable starting points, onboarding flows that guide product teams through the process of registering and configuring platform services for a new use case, and documentation that answers the questions product teams actually have, not the questions the platform team imagines they will have. All three require continuous refinement based on what product teams struggle with.

Diagram 7.3 – Self-Service Onboarding Flow

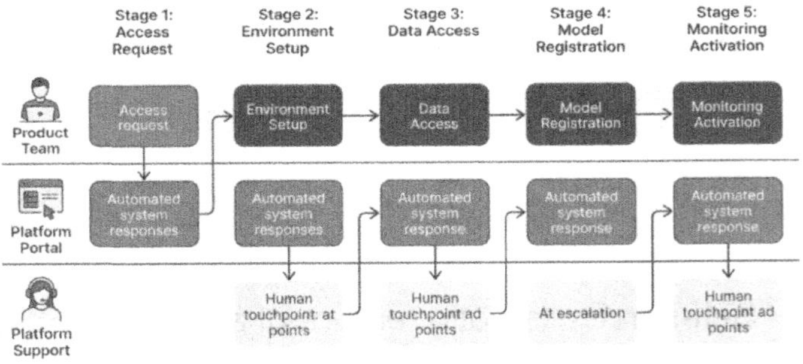

Self-service onboarding reduces platform team burden while giving product teams faster AI access.

Templates and Onboarding Flows

Templates are the platform's mechanism for encoding its own best practices. A deployment template for a classification model bundles the model registry integration, the monitoring hook configuration, the serving infrastructure specification, and the rollback mechanism into a single artifact that a product team can instantiate with their specific parameters. The product team does not need to understand all the implementation details; they need to understand what the template does and how to configure it for their use case.

Effective templates are opinionated but configurable. They make the common case trivially easy and the uncommon case possible. A template that forces every team to use identical monitoring thresholds is not a template; it is a constraint. A template that provides sensible defaults with documented override mechanisms gives product teams the starting point they need and the flexibility they will eventually require.

Onboarding flows complement templates by guiding product teams through the sequence of decisions they need to make when adopting platform services for the first time. A good onboarding flow is wizard-like: it surfaces one decision at a time, explains why each decision matters, provides sensible defaults, and records the team's choices for future reference. The output of an onboarding flow should be a configuration artifact that the team can control and modify over time, not just a set of settings buried in a portal interface.

Measure the effectiveness of templates and onboarding flows by tracking time-to-first deployment for new product teams. If that metric is not improving over time, the templates and flows are not doing their job. Interview teams that take longer than expected and identify where they get stuck. The platform team's job is to remove those friction points, not to provide more documentation about why friction exists.

Cost Quotas and Approval Thresholds

Cost management on AI platforms is not primarily a technical problem; it is a behavioral problem. Engineers experimenting with large models at inference time will consume significant compute without a feedback mechanism to inform them of the cost. Cost quotas provide feedback by creating a visible, enforced budget for each team's platform consumption. When a team approaches its quota, the platform surfaces a warning. When it reaches the quota, requests are throttled or blocked until the team reviews its consumption and either adjusts its approach or requests a quota increase.

The approval threshold system complements quotas by creating a tiered review process for different categories of platform resource consumption. Experiments below a defined cost threshold proceed automatically. Experiments above the threshold require a brief justification for the team's product manager to review. Production deployments above a higher threshold require a formal cost-benefit review that includes the platform team and, for significant expenditures, a finance stakeholder. This tiered approach keeps the governance burden proportional to the financial and operational risk.

Approval thresholds should also apply to non-cost decisions that carry organizational risk. A model that will affect clinical decisions, financial determinations, or legal outcomes should require a review that is not required for a model that will suggest document categories or prioritize support tickets. The platform can enforce these thresholds by requiring a use-case risk classification during onboarding and automatically applying the corresponding approval workflow. Governance becomes a property of the use case classification, not a judgment call made on a per-deployment basis.

Quotas and thresholds work only if they are set at levels that reflect real constraints rather than arbitrary conservatism. Quotas that are too tight lead to constant approval requests, which desensitize reviewers. Quotas that are too loose provide no meaningful signal. Calibrate quotas against actual usage from the first six months of platform operation, and revisit them quarterly as usage evolves. A quota system that is never revised has ceased to be managed.

Runbooks and Escalation Paths

A runbook is the platform team's contribution to the product team's operational capability. It is a documented, step-by-step response to a specific class of operational situations: what to do when a model's prediction latency spikes, when the feature store returns stale data, and when a monitoring alert fires and the on-call engineer is unsure of its meaning. Runbooks convert platform knowledge into product team capabilities, reducing the need to escalate to the platform team for routine operational situations.

Effective runbooks are written from the product team's perspective, not the platform team's. They should assume that the reader knows the use case well but may not know the platform's infrastructure in depth. They should link to the specific platform interfaces the reader will need to interact with, explain what each interface shows and what it means, and provide explicit decision criteria for each step. A runbook that says 'check the monitoring dashboard' is not useful. A runbook that says 'open the Model Health dashboard, select your model from the dropdown, and look for a sustained spike in the p95 latency metric above 500ms; if present, proceed to step 4' is useful.

The escalation path is the formal mechanism by which product teams contact platform engineers when runbooks are insufficient. Every platform should define a tiered escalation structure: self-service first, then async support with a defined response SLA, then a live escalation channel for production-impacting incidents. The SLAs at each tier should be published and tracked. A published but unmonitored platform that supports an SLA is a promise that will eventually be broken at the worst possible moment, damaging trust.

Platform teams should review escalation logs monthly to identify patterns. If the same question keeps surfacing in escalations, the runbook for that situation is inadequate or missing. If escalations consistently arrive from the same product team, that team may need more targeted onboarding support. If escalations consistently arrive at a particular point in the deployment workflow, the workflow has a friction point that the platform should resolve. Escalation data is the platform team's primary signal of where its self-service capability falls short.

7.6 Platform Governance and Funding

Platform governance encompasses the decisions about what the platform builds, who it serves, how it is funded, and what happens to capabilities that are no longer needed. These are organizational decisions as much as technical ones, and getting them wrong creates structural problems that technical excellence cannot fix. A platform team with excellent engineering but no clear governance model will struggle to prioritize, will underinvest in the services that matter most, and will eventually find itself either ignored by product teams or overwhelmed by competing demands.

The governance structure for an AI platform should reflect the seriousness of what it enables. If AI-assisted features influence clinical decisions, financial decisions, or public-facing services, the platform that enables them is a critical part of the organization's infrastructure. It should be governed accordingly, not with the informality appropriate to an experimental initiative, but with the clarity appropriate to infrastructure on which the organization depends.

Chargeback vs. Central Funding

The funding model for a platform team determines the incentives that govern its every decision. Under a chargeback model, the platform team recovers its costs by billing product teams for platform consumption. Under a central funding model, platform costs are absorbed by a central budget and distributed to product teams at no direct charge. Each model has significant organizational implications that go beyond accounting.

Chargeback creates cost transparency and accountability. Product teams can see exactly what they consume, which creates pressure to use platform resources efficiently. It also creates a revenue signal for the platform team: capabilities that product teams value enough to pay for are clearly worth maintaining; capabilities that no one uses, even if they are zero-cost, are clearly not. Chargeback also makes the platform team directly accountable to its customers in financial terms. If the platform is not delivering value, product teams will push back on the bills, which creates a productive feedback loop.

The risk of chargeback is that it can create perverse incentives during the adoption phase, when the platform needs product teams to experiment with it before its value is fully proven. If early adopters incur high costs for platform experimentation, they will be conservative in their use, preventing the platform from reaching the adoption threshold at which its benefits become clear. Many organizations address this by combining an adoption period of subsidized or free access with a chargeback model that phases in as the platform matures and usage stabilizes.

Central funding removes the adoption friction but creates a different problem: the platform team has no direct market signal from its customers. It must rely on stakeholder councils, usage metrics, and qualitative feedback to determine what to build. Without a financial feedback mechanism, platform teams can drift toward building impressive technical capabilities that product teams actually don't need. Central funding works best when it is paired with a strong stakeholder governance structure that gives product teams formal influence over platform priorities.

Diagram 7.4 – Funding Model Comparison

Dimension	Chargeback Model		Central Funding Model	
Cost Transparency	High	✓	Low	✗
Adoption Incentive	Low (teams avoid usage)	✗	High (usage encouraged)	✓
Platform Investment	Reactive	✗	Proactive	✓
Governance Leverage	Weak	✗	Strong	✓
Best Fit	Large distributed orgs		Orgs scaling AI	

Central funding drives AI adoption and platform investment; chargeback maximizes cost transparency.

Prioritization and the Stakeholder Council

Every platform team receives more requests than it can fulfill. Prioritization is not optional; it is the platform team's most consequential management function. The question is whether prioritization happens through a transparent, principled process or through informal influence, proximity to leadership, and whoever made the most noise in last week's meeting. The latter is common. The former is what distinguishes a well-run platform from a reactive one.

A stakeholder council is the governance mechanism that legitimizes platform prioritization. The council consists of representatives from the major product teams and business units that depend on the platform, as well as the platform team leaders and appropriate executives with an interest in AI governance. The council meets on a regular cadence, monthly or quarterly, to review the platform roadmap, surface new requirements from product teams, review adoption metrics, and make explicit prioritization decisions when competing requests conflict.

The council should operate with clear decision rights. The platform team should own decisions about how to build capabilities. The council should decide which capabilities to prioritize next, given the platform team's transparent set of trade-offs. Executives on the council should reserve their influence for decisions with significant organizational or financial implications, not for routine roadmap sequencing. Councils that lack clear decision-making rights become advisory bodies whose recommendations the platform team is free to ignore, thereby entirely defeating the purpose of governance.

Between council meetings, the platform team should maintain a public product backlog that shows what is being built, what is planned next, and what has been deferred, along with why. This transparency serves two purposes: it prevents product teams from making roadmap assumptions that later prove false, and it allows them to contribute their voices to prioritization discussions with full information about the current state of the backlog. Surprises in platform planning are an organizational cost. Transparency reduces them.

Deprecation Policy and Platform Evolution

Every capability a platform team builds creates a maintenance obligation. As the platform evolves, as better-serving infrastructure becomes available, as governance requirements change, and as product teams' needs shift, older capabilities need to be retired to keep the platform maintainable. Without a deprecation policy, the platform accumulates technical debt in the form of capabilities that are rarely used but expensive to maintain, APIs that are still active but no longer best practices, and integration patterns that newer product teams should not replicate.

A deprecation policy defines the process by which a platform capability moves from active to deprecated to retired. The process should include a minimum notice period, typically ninety days for non-critical capabilities, six months or more for capabilities on which multiple production systems depend, a migration guide that helps dependent teams move to the replacement capability, and a support commitment that specifies what assistance the platform team will provide during the migration window. The policy should also define what triggers a deprecation decision: readiness for replacement, usage below a threshold, a security concern, or a significant change in organizational direction.

Platform evolution also requires a maturity model that gives product teams a shared vocabulary for where the platform is and where it is headed. A four-level maturity model, foundational, standardized, optimized, and intelligent, helps stakeholders understand what the platform currently provides, what it is building toward, and what the benefits of each level look like in practice. The maturity model also helps the platform team communicate what investment is required to move from one level to the next, making the funding conversation more concrete.

One organizational pattern that helps manage platform evolution is a permanent pilot-to-production pathway for new platform capabilities before they are widely promoted. Before a new platform feature is included in templates and onboarding flows, it should be validated with two or three product teams in a controlled pilot. The pilot surfaces integration problems, documentation gaps, and UX issues that the platform team could not anticipate in development. Only capabilities that have survived should be presented to the broader product community as the recommended approach.

Diagram 7.5 – Platform Maturity Roadmap

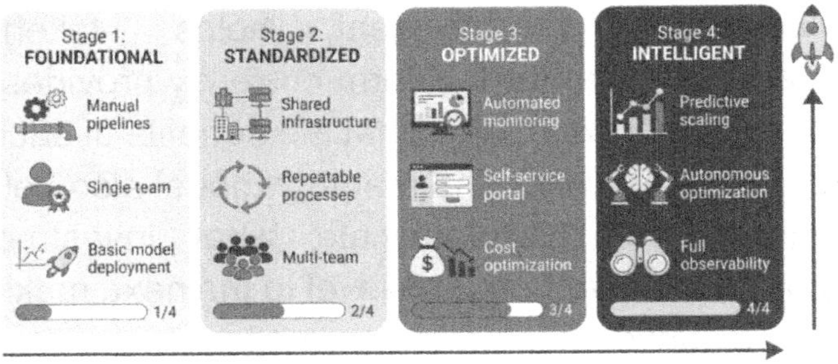

Platform maturity advances from foundational manual pipelines to intelligent autonomous optimization.

7.7 Manager's Checklist

Use this checklist to assess your AI platform's readiness and identify gaps that warrant immediate attention.

- Feature store in place with governance process for canonical feature definitions and versioning.
- Model registry is active and required as a hard prerequisite for platform-provided monitoring and serving.
- Monitoring hooks attached to all models in production, with product-team-configured alert thresholds.
- Documented product-platform boundary map that has been reviewed and accepted by both platform and product leadership.
- Self-service templates are available for the two or three most common AI deployment patterns in your organization.

- Onboarding flow documented and validated with at least two product teams who provided feedback.
- Cost quotas set and monitored, with tiered approval thresholds for spending above defined levels.
- Runbooks published for the ten most common operational situations product teams encounter.
- Support SLAs defined, published, and tracked for each tier of the escalation path.
- Stakeholder council chartered with clear decision rights, meeting cadence, and membership.
- Funding model defined with an explicit plan for the transition from early-adoption subsidy (if any) to steady-state funding.
- Deprecation policy written and communicated to all product teams currently consuming platform services.
- Platform maturity model visible to all stakeholders, with current status and next-stage investment requirements documented.
- Time-to-first-deployment metric tracked for new product teams and showing improvement over successive cohorts.
- Pilot-to-production pathway applied to all new platform capabilities before broad promotion.

7.8 Key Takeaway

An AI platform is the organizational infrastructure that enables an enterprise to convert individual AI experiments into a durable, scalable, governable capability. Its value is not in the technology it provides, but in the constraints it enforces, the standards it makes easy to follow, and the visibility it creates for managers who need to oversee AI at scale without being embedded in every product team's daily decisions.

The platform team's core design challenge is to maximize product team velocity while staying within boundaries that meet the organization's governance requirements. This requires clear thinking about what the platform owns versus what product teams own; sustained investment in self-service capabilities that reduce dependence on platform engineering time; and governance structures, funding models, stakeholder councils, and deprecation policies that make the platform sustainable over the multi-year horizon during which AI capabilities actually develop.

The health system in the opening scenario of this chapter did not succeed by building a more sophisticated platform. They succeeded by making the right decisions easier than the wrong ones, by designing a platform where the path of least resistance led to consistency, auditability, and reuse rather than to duplication and isolation. That design orientation, more than any technical choice, is what defines a platform that keeps product teams genuinely accountable.

8 Human-Centered Oversight and Scalable HITL Patterns

8.1 Scenario Scaling Safely with Tiered Review

The Director of Product at a federal benefits agency had a problem that appeared to be a success story from the outside. The agency deployed an AI-assisted eligibility pre-screening tool that processed 12,000 applications per day, a fivefold increase in throughput over the fully manual process it replaced, and processing times had dropped from 21 days to 4. Congressional stakeholders were satisfied. The agency had received favorable coverage in government technology publications.

Inside the agency, the picture was more complicated. The pre-screening model directed about forty percent of applications to human reviewers for final determination. Those reviewers were drowning. They were receiving hundreds of cases per day in a queue that provided no context for why each case had been flagged, no prioritization by urgency or risk, and no guidance on which aspects of the case required the most careful attention. Reviewers worked through cases as quickly as they could, so they often did not review them as carefully as they should. Approval error rates had crept up over six months. Three cases had resulted in formal appeals that cited inadequate review.

The Director brought together her product, operations, and policy teams to redesign the human-in-the-loop component. They implemented a confidence-tiered review system: high-confidence cases received a lightweight spot-check review by one reviewer, medium-confidence cases received a standard full review, and low-confidence cases received an enhanced dual review with mandatory supervisor sign-off. They redesigned the reviewer interface to show each case's confidence tier, the specific factors that had triggered the flag, and the relevant policy criteria for that case type. They added inter-annotator agreement tracking to identify cases where two reviewers reached different conclusions, which fed directly into model retraining. Within two quarters, review quality had improved, appeal rates had dropped, and the average time per case had decreased because reviewers were spending their attention where it was most warranted. The cost per correct determination was down by thirty-one percent.

8.2 Why Human Oversight Is a Design Variable

The decision about how humans are involved in an AI-assisted workflow is not a policy-compliance question; it is a product-design question. Human-in-the-loop, or HITL, refers to any design in which a human reviewer plays a role in evaluating, approving, correcting, or escalating model outputs before or after they affect a consequential decision. The 'in the loop' framing is useful because it emphasizes that the relationship between human and model is a system, not just an approval layer bolted onto an otherwise automated process.

Organizations frequently treat human oversight as a fixed cost to be minimized. This framing is both strategically and operationally wrong. Human oversight is a design variable with tradeoffs. More human involvement improves review quality and reduces the risk of systematic errors, but also increases costs and processing time. Less human involvement reduces costs and increases throughput but concentrates risk in cases where the model is least reliable. The right design is not maximum automation or maximum review; it is a calibrated allocation of human attention to the decisions where it adds the most value.

The calibration problem is harder than it sounds. It requires knowing which model outputs are most likely to be wrong, which errors are most costly, and how much human attention is needed to catch the errors that matter. These are empirical questions that require data and ongoing measurement. Organizations that design HITL workflows without that data are making guesses, and their guesses often lead to over-reviewing low-risk cases and under-reviewing high-risk ones.

This chapter provides a framework for designing HITL workflows as a deliberate system with measurable outcomes. It addresses the pattern design confidence thresholds, selective review, escalation queues, the reviewer experience, the measurement approach, and the operational infrastructure for incident management and cost control. The audience is the manager who oversees the product, operations, or quality function for an AI-assisted process, the person who is responsible for outcomes, not just for deploying technology.

Responsible by design in the HITL context means building a review system that makes it easy for humans to do their job well, provides feedback loops to improve both the model and the human process over time, and creates clear accountability for the quality of outcomes. It does not mean maximizing human review for its own sake, nor does it mean minimizing it in the name of efficiency. It means designing the system to achieve the best results given the available resources and the use case's risk profile.

8.3 Scalable HITL Patterns

Three patterns cover the majority of HITL implementations in enterprise AI systems: confidence-threshold review, selective review, and escalation queues. These patterns are not mutually exclusive; a well-designed HITL workflow will typically combine all three, applying each to different categories of model outputs based on the outputs' risk and confidence characteristics.

The choice of HITL pattern is a strategic decision, not a tactical one. It determines the total cost of human review as a function of volume, the risk exposure the organization accepts from automated decisions, and the quality signal the model receives from human corrections. Managers who leave HITL pattern design to engineering teams often end up with technically functional but strategically miscalibrated systems in which humans review the wrong cases, provide feedback in the wrong format, or operate under conditions that prevent them from maintaining quality over time.

Diagram 8.1 – Tiered HITL Decision Flow

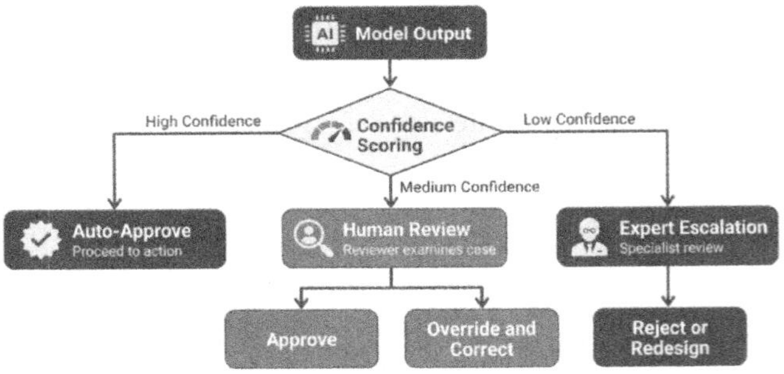

Route AI model outputs through three confidence tiers to appropriate human review levels.

Confidence Thresholds

A confidence threshold is a model-output score below which a human review is automatically triggered. The threshold is expressed as a probability or score that reflects the model's own estimate of certainty about its output. When the model is highly confident, when the patterns in the input strongly resemble the patterns it was trained to recognize, the output can be accepted with minimal review. When the model is less certain, human attention is warranted.

Setting confidence thresholds requires calibration against actual outcome data. The threshold should be set such that the error rate among cases that clear it, that is, cases the model handles without triggering review, is acceptable given the cost of errors in that use case. For a document categorization system, an error rate of 2 or 3% among auto-approved cases may be entirely acceptable. For a clinical risk-stratification system, the threshold may need to be set to keep the auto-approved error rate near zero, even if that means routing a larger fraction of cases to human review.

Confidence thresholds are not static. They should be recalibrated regularly as the model is retrained, as the input data distribution shifts, and as the organization's risk tolerance evolves. A threshold that was appropriate six months ago may be allowing too many errors through today if the input population has changed in ways the current model was not trained to handle. Monitoring the actual error rate among auto-approved cases, not just the model's confidence score, is the metric that tells you whether the threshold is still correctly calibrated.

One practical complexity: many production models generate confidence scores that are not well-calibrated to actual error rates. A model that reports 90% confidence may be correct on its 90% confidence predictions only 70% of the time. Calibration evaluation, comparing confidence scores to actual outcomes on a held-out set, should be a standard part of model evaluation before deployment, and recalibration should be included in the retraining process. Using raw confidence scores as thresholds without calibration validation is a common source of systematic failure in HITL design.

Selective Review

Selective review extends the confidence-threshold concept to incorporate additional signals beyond raw model confidence. Rather than routing cases for review solely based on the model's self-reported uncertainty, a selective review system uses a review propensity model, a secondary model, or a scoring rule to estimate the probability that human review will change the outcome for a given case. This allows the review queue to be prioritized by expected value rather than by raw volume.

The distinction is important. A high-confidence case that happens to fall into a category where the model has historically been wrong more often than its confidence scores suggest, for example, cases involving a recently changed policy, a specific demographic subgroup, or an unusual data pattern, may warrant review even though it clears the confidence threshold. Conversely, a moderate-confidence case in a category where human reviewers almost always agree with the model may not be worth the reviewer's time despite its lower confidence score.

Selective review requires a feedback infrastructure that the simpler confidence-threshold approach does not. To train a review propensity model, the organization needs labeled data indicating which cases were reviewed, what the model predicted, what the reviewer concluded, and whether the reviewer agreed. This data accumulates naturally in a well-instrumented HITL workflow. Still, it must be collected deliberately and linked to case outcomes in a way that makes it usable for propensity modeling. Organizations that do not invest in this instrumentation cannot implement selective review effectively.

The strategic value of selective review lies in cost efficiency without sacrificing quality. By directing human attention to the cases that matter most, the organization can maintain review quality while reducing total review volume or maintain volume while improving quality. In the federal agency scenario that opens this chapter, selective review was the mechanism that allowed the agency to reduce the average review time per case while improving overall accuracy: reviewers spent their time on cases that warranted careful attention, not on cases where the model's output was nearly certain to be correct.

Escalation Queues

An escalation queue is a specialized review pathway for cases that meet specific criteria requiring a higher level of review authority or expertise than the standard reviewer population can provide. Escalation criteria might include cases involving amounts above a financial threshold, cases in which a previous reviewer's decision has been appealed, cases in a domain or subpopulation that require specialized knowledge, or cases in which the model's output falls within a range that has historically exhibited elevated error rates.

Escalation queues serve two functions simultaneously: they route complex cases to the reviewers best equipped to handle them, and they prevent those cases from creating bottlenecks in the standard review queue. Without a formal escalation queue, complex cases languish in the standard queue. At the same time, reviewers try to figure out what to do with them, or they get pushed through the standard process by reviewers who lack the authority or expertise to handle them correctly.

Managing escalation queues requires careful attention to queue depth and throughput. An escalation queue that is growing faster than it is being cleared signals that either the escalation criteria are too broad, too many cases are being escalated, or the escalation reviewer's capacity is insufficient. Either problem, left unaddressed, creates a backlog that eventually pressures escalation reviewers to rush through cases, which defeats the purpose of escalation. Escalation queue health should be a standing item in any HITL operational review.

8.4 Designing Reviewer Experience

The reviewer's experience is part of the product design layer at HITL. It encompasses every aspect of how a human reviewer interacts with a case: what information they see, in what order, with what level of context, and through what interface. Poor reviewer experience is the most common reason that technically sound HITL systems fail to deliver quality outcomes. A reviewer who cannot quickly understand what they are being asked to evaluate, or who must navigate a confusing interface to record their decision, will make faster but less careful decisions, meaning the human oversight that justified the HITL design is not actually occurring.

Reviewer experience design is not a UX problem that an engineer can solve by making the interface cleaner. It requires understanding the cognitive demands of the review task, the information that reviewers need to make a good decision, and the pressures, such as throughput targets, queue depth, and colleague dynamics, that influence reviewer behavior in practice. This understanding comes from spending time with reviewers as they do their work, not from designing an interface in isolation and measuring throughput metrics from a distance.

Diagram 8.2 – Reviewer Experience Design

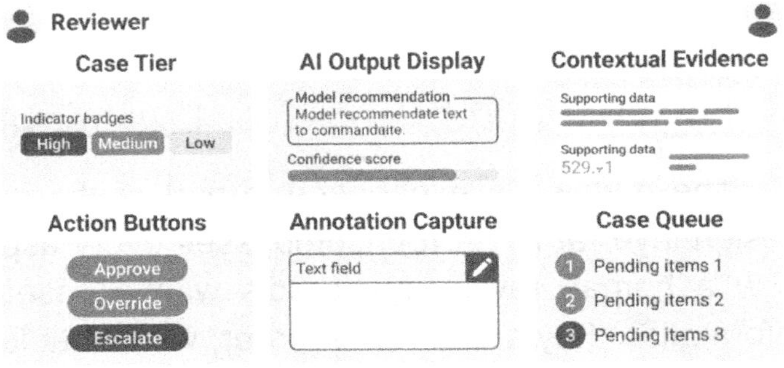

A well-designed reviewer interface reduces cognitive load and improves annotation quality.

Batching and Context Windows

Batching is the practice of presenting related cases together in a reviewer's queue, rather than serving them in purely random order. Thoughtful batching reduces the cognitive switching cost that reviewers incur when each case requires a completely different mental model. A reviewer who reviews ten cases involving the same policy provision in sequence will make more consistent decisions than a reviewer who alternates between ten different case types, because they can build and apply a consistent mental model rather than context-switching with each case.

Context windows are the temporal scope within which a reviewer's decisions are presented back to them for self-monitoring. A context window shows the reviewer their last 20 decisions, including the outcomes and any inter-annotator disagreements, so they can identify whether their decision pattern has drifted, whether they have been approving more or fewer cases than usual, or if their decisions are systematically diverging from those of other reviewers on similar cases. Context windows support reviewer self-calibration, an important quality-maintenance mechanism, especially in high-volume queues where individual decisions can blur together.

The combination of batching and context windows addresses one of the most persistent quality problems in human review: reviewer drift. Reviewers who work in high-volume queues for extended periods tend to develop idiosyncratic decision-making patterns that diverge from both the written criteria and their colleagues' criteria. This drift is not intentional; it is a natural consequence of the human tendency to develop heuristics that reduce cognitive load. Batching and context windows help reviewers notice and correct drift before it becomes a systematic quality problem.

Minimizing Cognitive Load

Cognitive load is the total mental effort required to evaluate a case and record a decision. When cognitive load is high because the interface is complex, the case presentation is disorganized, or the decision criteria are ambiguous, reviewer accuracy declines. The relationship between cognitive load and accuracy is well established in the human factors literature. For managers, the implication is straightforward: every element of the reviewer interface that requires mental effort is a potential source of decision error.

Minimizing cognitive load starts by surfacing only the information relevant to the decision at hand. A reviewer evaluating a prior authorization claim does not need to see the full medical record; they need only the specific data points referenced by the policy criteria. A reviewer evaluating a loan application does not need to see the full credit file; they need to see the summary signals that the model identified as influential and the policy criteria that apply to this case type. More information is not better. Precisely relevant information is better.

Decision scaffolding, structured decision templates that guide the reviewer through the evaluation criteria in a logical sequence, reduces cognitive load by providing a clear process rather than requiring the reviewer to reconstruct the evaluation criteria from memory. Effective scaffolding does not force reviewers into a rigid checklist that prevents them from exercising judgment. It provides a structure that enables consistent application of judgment across a high volume of cases.

Reviewer interface design should be treated as an ongoing product function, not a one-time implementation. The right design for a queue of fifty cases per day differs from that for a queue of five hundred. The right design for a case type that reviewers handle daily differs from the right design for one that appears only once a month. As volume scales and the HITL system matures, the reviewer interface should evolve alongside it, informed by direct observation of reviewer behavior and systematic analysis of decision-quality metrics.

Throughput and Quality Gates

Throughput metrics measure how many cases a reviewer processes over a given period. Quality gates are threshold standards that a reviewer's decision quality must meet to remain in the review function. Both metrics matter, and both can be gamed if they are not designed carefully. A throughput target without a quality gate creates pressure to process cases quickly without adequate care. A quality gate without a throughput metric creates no incentive to work efficiently.

The most effective throughput and quality metric systems set a minimum acceptable quality level as a hard constraint and then optimize for throughput within that constraint. A reviewer who meets quality standards and processes cases efficiently is doing exactly what the organization needs. A reviewer who meets quality standards, but processes cases very slowly is a capacity constraint. A reviewer who processes cases quickly but fails quality standards is a risk. Management responses to each situation differ, and a combined metric system makes it possible to distinguish between them.

Quality gates should be based on inter-annotator agreement, the rate at which a reviewer's decisions align with those of other qualified reviewers on the same cases, rather than on agreement with the model. Using model agreement as a quality metric creates a circular feedback loop: reviewers who agree with the model appear to be performing well, even if the model is wrong. Inter-annotator agreement measures whether reviewers apply the decision criteria consistently, which is the actual quality standard the organization cares about.

Quality gates also serve an important organizational function: they make the expectations of the review role explicit and auditable. In regulated industries, the ability to demonstrate that review decisions were made by reviewers who met defined quality standards is often a compliance requirement. A documented, measured, and consistently enforced quality gate system is the evidentiary foundation for that demonstration.

8.5 Measuring HITL Effectiveness

A HITL system that is not measured cannot be managed. The organization may believe that human review is improving outcomes, and it may be right. Still, without measurement, there is no way to know whether the HITL design is calibrated correctly, whether it is improving over time, or whether the costs it imposes are justified by the benefits it delivers. Measurement is what converts HITL from an act of faith into an engineering discipline.

HITL measurement spans three dimensions: input quality (how useful is the data the reviewers are generating?), process quality (is the review process producing consistent, reliable decisions?), and outcome quality (is the combination of model and human review producing better outcomes than either alone would?). Organizations that measure only one dimension will optimize for it at the expense of the others. An organization that focuses only on throughput will sacrifice quality. An organization that focuses solely on inter-annotator agreement may overlook the fact that reviewers consistently agree on wrong answers because the decision criteria are poorly designed.

Diagram 8.3 – HITL Measurement Framework

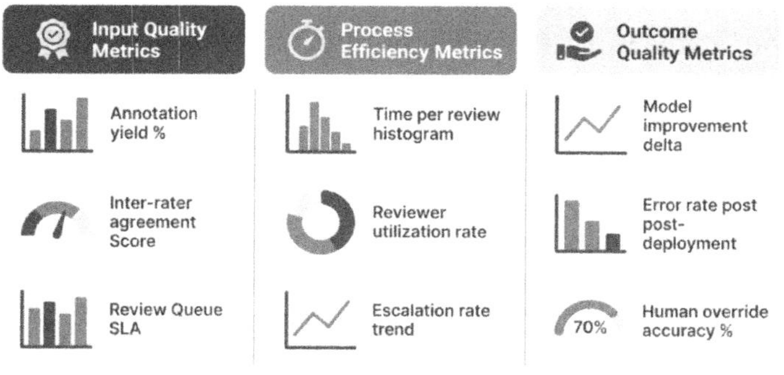

Measure HITL system health across input quality, process efficiency, and downstream outcome quality.

Annotation Yield and Cost per Correction

Annotation yield measures the fraction of reviewed cases that yield a label actually useful for model improvement. Not every reviewed case is informative. In cases where the model was already correct, and the reviewer confirms the model's output, add no training signal; such confirmations are expensive confirmations of the model's existing knowledge. Cases where the reviewer disagrees with the model, especially cases where the disagreement reflects a genuine pattern the model should learn, are the high-yield cases that make HITL investment worthwhile.

Tracking the yield of annotations over time reveals whether the HITL system is generating diminishing returns. Early in a model's deployment, many cases will be informative because the model has not yet encountered the full diversity of real-world inputs. Over time, as the model is retrained on reviewer feedback, the fraction of reviewed cases that the model already handles correctly will increase, and annotation yield will decline. The declining yield is not necessarily bad; it may indicate that the model has improved, but it should trigger a reconsideration of HITL design. Is the current review rate still justified by the yield it produces? Would a higher confidence threshold, routing fewer cases to review, maintain quality while reducing cost?

Cost per correction measures the total cost of human review, reviewer time, tooling, and management overhead divided by the number of cases in which the review changed or improved the outcome. It is the fundamental efficiency metric for HITL. An organization that spends $4,000 per correction in a process where each error costs $200 in downstream rework has a HITL system that is not cost-justified, regardless of its technical sophistication. An organization that is spending $200 per correction in a process where each error costs $50,000 in downstream liability has a HITL system that is almost certainly underinvested.

Cost per correction should be computed separately for each tier of the review system. High-confidence spot-check reviews have a very low cost per review but also a low correction rate; most of those cases did not require review. Low-confidence enhanced reviews have a higher cost per review but a much higher correction rate. The cost per correction for the enhanced review tier is often lower than for the spot-check tier, even though the per-case cost is higher, because the correction rate is so much better. This counterintuitive result is one reason why tiered review systems are more efficient than uniform review systems at any volume.

Model Improvement per Annotation

Model improvement per annotation is the metric that closes the feedback loop between HITL operations and model development. It measures how much the model's performance improves on held-out evaluation data for each batch of reviewer annotations incorporated during retraining. Tracking this metric over successive retraining cycles allows the organization to observe the learning curve: how much improvement the model can still derive from reviewer feedback, and whether that improvement rate is declining.

The improvement metric should be broken down by case type or domain, not just measured in aggregate. A model might be reaching peak performance across most of its input distribution while still showing significant improvement potential for specific subpopulations or edge cases. An aggregate improvement rate that appears to plateau may be masking continued high-yield improvement opportunities in specific segments. Segment-level improvement tracking is the tool that reveals those opportunities.

When model improvement per annotation falls to near zero across all segments, the HITL system has served its primary purpose: improving the model. At that point, the HITL design should shift from improvement-oriented to quality-assurance-oriented: the role of human review is no longer to teach the model new things, but to catch cases where the model fails in ways that its self-reported confidence does not reveal. The design implications are significant; the review interface, the reviewer profile, and the throughput targets appropriate for a quality-assurance

HITL function differ from those appropriate for an improvement-oriented one.

Staffing and Forecasting

HITL staffing is a capacity-planning problem with several variables that are not fully within the organization's control. The review volume is a function of model confidence, input volume, and escalation rates. The time per case is a function of case complexity, interface design, and reviewer experience. The quality standard determines the amount of reviewer time required per case. None of these variables are perfectly predictable, and all can change rapidly in response to model retraining, policy changes, or shifts in the input distribution.

Effective HITL staffing requires a forecasting model that links model confidence metrics to projected review volume and translates that volume into required staffing levels, given the organization's throughput standards. This model should be updated as model performance evolves, especially after retraining events that change the confidence distribution. A model that becomes significantly more confident after retraining, which typically happens, will route fewer cases to review, and the staffing level calibrated to the pre-retrain volume will be excessive. Catching that mismatch before it results in idle reviewer capacity requires active forecasting.

Training new reviewers is a significant time and cost investment that must be factored into staffing plans. A reviewer new to the task will work more slowly and have a lower initial quality score than an experienced reviewer. Building a staffing plan that accounts for the ramp-up period, the time it takes a new reviewer to reach the throughput and quality standards of an experienced one, prevents the common mistake of hiring to exactly projected demand and then finding that the new staff cannot deliver at projected capacity for the first four to eight weeks.

Reviewer retention is a material operational risk for HITL systems, often underestimated. High-volume review roles are cognitively demanding, often repetitive, and can feel disconnected from the outcomes the organization cares about. Organizations that do not invest in making the reviewer role meaningful through feedback on outcomes, visibility into how their decisions affect results, and a clear pathway for skilled reviewers to take on more complex responsibilities will experience high turnover in the reviewer function. The cost of that turnover, in recruiting and training, often exceeds the cost of the retention investments that would have prevented it.

8.6 Escalation and Incident Handling

Even a well-designed HITL system will encounter situations outside the envelope it was designed to handle. A sudden change in the characteristics of the input data can result in a flood of low-confidence cases that overwhelm the review queue. A policy change can invalidate the decision criteria the reviewers are applying. A discovered systematic error may require retroactive review of a large case cohort. An external incident, a regulator inquiry, a media report, or a high-profile appeal can require a rapid operational response at a time when the team is already operating at capacity.

Incident handling is the operational capability that allows a HITL system to absorb these events without losing control. It requires prepared playbooks that define the response protocol for the most likely incident types, clear escalation paths that identify who is responsible for each decision class, and communication templates that enable the team to notify stakeholders quickly and accurately without causing additional confusion. Organizations that build these capabilities before they need them respond effectively to incidents. Organizations that try to build them in the middle of an incident do not.

Diagram 8.4 – Escalation and Incident Flow

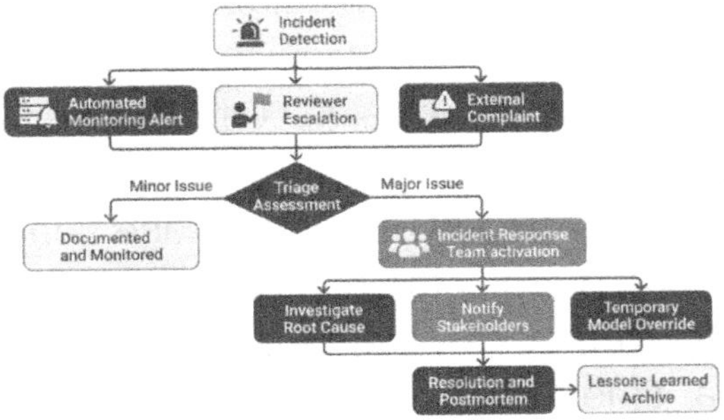

A structured escalation flow ensures fast triage and systematic resolution of AI system incidents.

Triage Flows and Retrain Triggers

A triage flow is a structured decision process that transforms an ambiguous situation- something is wrong, but it is not yet clear what or how badly- into a classified incident with a defined response protocol. The triage process should answer four questions in rapid sequence: What is the scope? How many cases are affected and over what period? What is the likely cause? Data shift, model degradation, process change, or external event? What is the risk? What is the potential harm if the situation is not addressed immediately? And what is the response? What actions are available, who can authorize them, and what are the tradeoffs?

Retrain triggers are pre-defined conditions that automatically place a model on the retraining schedule, flagging that its performance has degraded enough to warrant an update. Defining retrain triggers in advance is important because it prevents the retrain decision from becoming a political negotiation during an operational incident. If the trigger condition is met- error rate above threshold, data drift index above threshold, or inter-annotator agreement below threshold- the retrain is initiated. The decision was made when the policy was written, not when the pressure is highest.

Common retrain triggers include: error rate on a rolling review sample exceeding a defined threshold, input feature distribution shifting by more than a defined amount from the training distribution, inter-annotator agreement dropping below the quality gate threshold for more than one week, or a calendar-based trigger that ensures models are retrained at a minimum frequency regardless of performance metrics. The calendar trigger is the most basic and should be a floor, not a ceiling. A model that meets all performance thresholds but has not been retrained in the past 18 months is almost certainly missing improvements that more recent data would reveal.

Communication Templates and Rollback Criteria

Communication templates serve as a critical function during incidents: they allow the team to notify stakeholders accurately and quickly, without requiring someone to compose an incident notification from scratch while managing the operational response. A well-designed template covers the incident summary, the affected population, the actions taken to date, the next steps with owners and timelines, and the criteria that indicate resolution. Templates should be version-controlled and reviewed periodically, not just pulled out during incidents.

Rollback criteria define the conditions under which a model deployment should be reversed: the organization should stop routing new cases to the current model version and revert to a previous version or to a fully manual process. Rollback is not a comfortable decision. It acknowledges that the current model is not performing adequately, which may generate internal and external scrutiny. Organizations that do not define rollback criteria in advance will resist rolling back even when it is the right operational decision, because the criteria for 'bad enough' are subjective, and the decision is under pressure.

Defined rollback criteria remove the subjectivity and the pressure. If the error rate on reviewed cases exceeds a defined threshold for more than four consecutive hours, rollback is initiated. The decision is based on data, not on judgment under pressure. The team still needs to execute the rollback and communicate it to stakeholders. But the decision itself is taken off the table because it was made in advance, under calmer conditions, by people with the appropriate authority.

Rollback criteria should be part of the production approval process for every model deployment, not a capability that the team hopes it never needs. The governance workflow enforced by the model registry, as described in the previous chapter, should require that rollback criteria be specified and reviewed before a model is approved for production. This makes rollback preparedness a standard of operational maturity rather than an optional addition.

Cost Control and SLA Design

Cost control in HITL systems requires managing the two largest cost drivers: reviewer time and model-serving compute. Reviewer time costs are primarily determined by review volume, time per case, and reviewer compensation. Review volume is controlled by a confidence threshold design. The reviewer's design experience controls the time per case. Reviewer compensation is a staffing decision influenced by the labor market, the complexity of the review task, and the required quality standards. Of these three levers, confidence threshold design typically has the greatest cost impact and is the most actionable for a product manager.

Throttling is the operational mechanism that limits review volume when it would otherwise exceed capacity or budget. Rather than allowing the review queue to grow unboundedly during volume spikes, throttling pauses new case intake when the queue depth exceeds a defined threshold. Throttled cases are held in a waiting state rather than being auto-approved or dropped. The organization accepts a processing delay rather than a compromise in quality or an unbounded cost spike. Throttling requires that the upstream system, the channel through which new cases arrive, be able to hold cases in a waiting state. This infrastructure requirement should be designed in from the beginning.

SLA design determines what the organization promises stakeholders about HITL processing times. An SLA that is too aggressive, promising a twenty-four-hour review on cases that typically take forty-eight hours during volume peaks, will be regularly missed, eroding stakeholder confidence. An overly conservative SLA will put pressure on stakeholders who experience unnecessarily long waits. The right SLA is calibrated to the ninety-fifth-percentile processing time under normal operating conditions, with a separate SLA for expedited cases and a defined process for requesting expedited review.

Diagram 8.5 – HITL Cost Control Dashboard

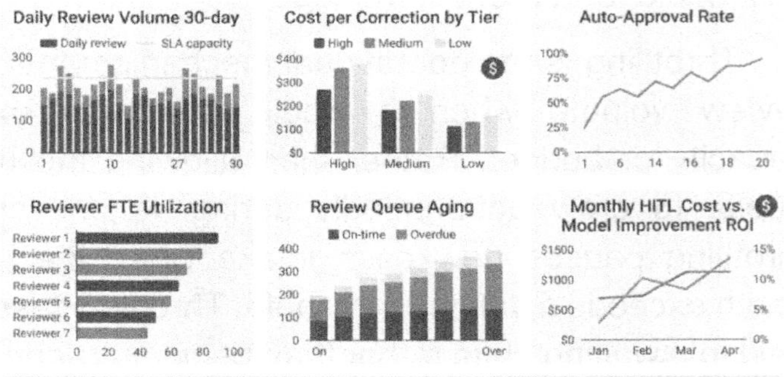

Six HITL cost control metrics help managers balance review quality with operational efficiency.

8.7 Manager's Checklist

Use this checklist to assess whether your HITL system is designed, measured, and managed to deliver reliable, cost-justified outcomes.

- Confidence thresholds set based on calibrated outcome data, not raw model confidence scores.
- Tiered review design in place with defined review protocols for each confidence tier.
- Selective review logic implemented or on the roadmap, using a review propensity model rather than raw confidence alone.
- Escalation queue defined with explicit escalation criteria, reviewer authority levels, and queue health monitoring.
- Reviewer interface tested with actual reviewers and revised based on observed behavior, not just self-reported satisfaction.
- Batching and context window mechanisms in place to support reviewer self-calibration.
- Throughput and quality gate metrics defined, tracked, and used in reviewer performance management.
- Annotation yield is tracked and reviewed at least monthly to assess whether the current review volume is generating a useful training signal.
- Cost per correction calculated by tier and reviewed against the downstream cost of errors.
- Model improvement per annotation tracked across retraining cycles and by case segment.

- HITL staffing forecast linked to model confidence distribution, updated after each retraining cycle.
- Reviewer training program, including ramp-up timeline and quality gate milestones for new reviewers.
- Retrain triggers defined, documented, and reviewed as part of the model production approval process.
- Rollback criteria documented for every model in production, with rollback procedure tested in a non-production environment.
- Communication templates prepared for the two or three most likely incident types in each case.
- Throttling mechanism in place to manage review volume spikes without compromising quality or accepting unbounded cost.
- HITL SLAs are defined at the ninety-fifth-percentile processing time, with a separate expedited SLA and request process.

8.8 Key Takeaway

Human-in-the-loop design is not a regulatory compliance mechanism dressed up as a product feature. It is a strategic capability that, when well designed, makes an AI-assisted process simultaneously more accurate, more cost-efficient, and more trustworthy than either full automation or full manual review could achieve on its own. The federal agency in this chapter's opening scenario did not reduce errors by hiring more reviewers; they reduced errors by giving reviewers the right cases to review, the right context to review them with, and a feedback loop that turned their decisions into systematic improvement.

The manager's job in HITL design is to hold the tension between scale and quality. Scale without quality is automation that erodes public trust and creates downstream liability. Quality without scale is a manual process with a machine in front of it, offering the costs of AI without its benefits. The patterns in this chapter- tiered confidence review, selective review, escalation queues, measurement frameworks, and cost controls- are the engineering discipline that allows scale and quality to coexist.

That discipline requires continuous attention. The inputs change as the model is retrained. The volume changes as the use case scales. The risk profile changes as the organization's regulatory environment evolves. A well-designed HITL system will drift out of calibration without active management. Mission-aligned oversight means keeping the human review system as current as the model it supports and recognizing that the two are not separate concerns but components of a single, integrated system whose quality is determined by how well they work together.

9 Measuring Impact and Making Change Durable

9.1 Scenario: The Portfolio Review That Saved Millions

The CIO of a mid-market property and casualty insurer had spent three years building an AI capability that, by any reasonable definition, had worked. The organization had deployed sixteen AI-assisted features across underwriting, claims, and customer service. Model performance metrics looked strong. The engineering team was running smoothly. The CIO received positive recognition at industry conferences. But when the CFO asked, during a board strategy session, what the AI program had returned in economic value, the CIO could not answer with any precision. She knew models were running. She did not know what they had changed.

The problem was attribution. Each AI feature had been deployed without a measurement architecture that connected model outputs to business outcomes. The underwriting recommendation model had been in production for eighteen months. Still, no one had tracked whether underwriters who used it wrote more profitable policies than those who did not. The claims triage model had been deployed for two years, but no one had measured whether it had reduced the average time from first notice of loss to payment decision. The customer service routing model had been running for a year, but no one had tracked whether it had reduced handle time, improved resolution rates, or affected renewal likelihood.

The CIO commissioned a portfolio review. She assembled a cross-functional team, including finance, operations, data science, and business unit leadership, and spent eight weeks building retrospective outcome analyses for each of the sixteen deployments. The results were illuminating. Six features delivered clear, measurable business value. Five were contributing to process improvement, but at a cost that exceeded the value they delivered. Four were essentially neutral; they had not made things meaningfully better or worse, which meant the organization was paying ongoing maintenance costs for no return. One had degraded outcomes in a specific product line, undetected because no one was measuring the right thing.

The CIO used these findings to reallocate investment. She sunset the neutral and underperforming features, redirected the freed capacity to two high-potential use cases that had been waiting for engineering resources, and implemented a mandatory measurement architecture that every new AI feature had to satisfy before it could be approved for production. The reallocation freed eleven percent of the total AI program budget. The two new features that received the redirected investment became the organization's two strongest performers within twelve months. The lesson was not that the AI program had failed; it was that, without portfolio discipline, even successful programs leave enormous value on the table.

9.2 Measurement Is the Foundation of Durable Chg.

An AI initiative that cannot demonstrate its impact is not a durable strategic asset; it is an expenditure awaiting justification. The difference between AI programs that compound in value over the years and those that stagnate or are cut when budget pressure arises is rarely due to model quality. It is the quality of the measurement architecture and the discipline with which the organization uses measurement to direct investment, retire underperforming capabilities, and compound on proven success.

Measurement serves multiple organizational functions simultaneously. For the product and engineering teams, measurement is a feedback mechanism: it tells them whether what they built is working as intended and what they should do differently. For business unit leaders, measurement serves as a justification mechanism: it provides the evidence they need to continue investing in AI-assisted operations and to defend that investment to finance and executive leadership. For the CIO or CAIO, measurement is a portfolio management tool: it provides the basis for reallocating investment from lower-value to higher-value uses of AI capability.

The challenge is that measurement is harder than deployment. Deploying a model requires a defined engineering process that, when executed on a mature platform, can be completed within a predictable timeline. Measuring the model's impact requires answering counterfactual questions, such as what would have happened without the model? That cannot always be answered cleanly, especially when the model was deployed without a controlled experiment. Attribution is genuinely difficult in complex organizational workflows, and the measurement architecture that makes it tractable must be designed before deployment, not retrofitted afterward.

This chapter provides a framework for building that measurement architecture. It addresses the minimal impact dashboard, the smallest set of metrics that gives managers meaningful visibility into AI portfolio performance, the executive view that translates technical metrics into business language, the portfolio management process that turns measurement into investment decisions, and the institutional practices that make learning and accountability durable over time.

The audience for this chapter is not the data scientist who builds the model or the engineer who deploys it. It is the manager who is accountable for outcomes, the VP of Underwriting, who owns the business results the AI feature is supposed to improve; the Chief Claims Officer, who has to answer for cycle-time metrics; or the CIO, who owns the AI program's strategic narrative to the board. For these leaders, measurement is the management instrument that links AI investment to the organizational performance it is intended to drive.

9.3 The Minimal Impact Dashboard

The minimal-impact dashboard is not the largest possible; it is the smallest that provides managers with the information they actually need to manage the portfolio. Most AI programs generate too many metrics, not too few. Teams track model accuracy, precision, recall, F1 score, data freshness, prediction latency, serving cost, feature importance, and dozens of other technical indicators. Most of those indicators are useful for the engineering team. Almost none of them are the right instruments for a manager who needs to decide whether to continue, expand, reduce, or sunset a given AI capability.

The minimal-impact dashboard is organized into four metric categories: business KPIs, model health metrics, learning metrics, and attribution links. Each category serves a distinct management purpose, and the discipline of limiting the dashboard to these four categories forces the organization to be explicit about what it actually cares about rather than collecting every technically available metric.

Diagram 9.1 – Minimal Impact Dashboard Layout

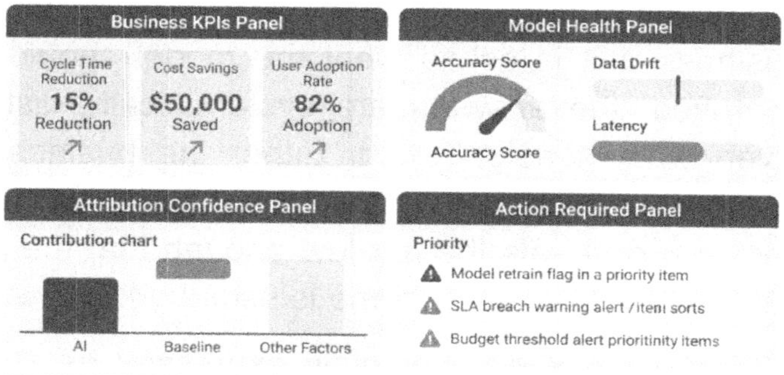

A minimal impact dashboard surfaces business value, model health, attribution confidence, and required actions.

Business KPIs

Business KPIs are metrics that measure whether the AI feature is achieving the organizational outcome for which it was deployed. They are defined by the business unit, not by the data science team, and expressed in the language of business operations rather than model performance. A claims triage model is measured by its impact on first-notice-to-decision cycle time and on claims leakage, not by its classification accuracy on the validation set. An underwriting recommendation model is measured by its impact on loss ratio in the policies where it was used and on underwriter productivity, not by its F1 score.

Defining the right business KPIs before deployment is one of the most important governance decisions an organization makes about an AI feature. It forces the business unit and the product team to agree in writing on what success looks like before anyone has a stake in the answer. When this conversation happens after deployment, and the business unit is asked what the model should have improved, the answers are systematically distorted, given what it actually did. Either the business unit claims the model improved things it did not actually improve, or it sets the bar low enough that the model can clear it. Neither outcome is useful.

Business KPIs should be expressed as specific, time-bounded, directional targets. Not 'improve claims cycle time' but 'reduce average first-notice-to-decision cycle time by fifteen percent within twelve months of deployment, measured in the product lines where the model is in use.' Specificity creates accountability. It makes it possible to tell, at the twelve-month mark, whether the target was reached and whether the deployment met its intended purpose. Vague KPIs create the appearance of accountability without its substance.

Not every business KPI can be measured with a clean before-and-after comparison. In workflows where AI assistance is applied to all cases simultaneously, with no control group, attribution requires either a quasi-experimental design, a comparison against a historical baseline with appropriate controls, or a modeled counterfactual. Each approach has limitations. The organization's job is to choose the most credible attribution approach available, given the deployment context, and to be explicit about its limitations in the reporting. Credible approximate attribution is more valuable than precise measurement of the wrong thing.

Model Health Metrics

Model health metrics indicate whether the model is currently operating within the performance envelope against which it was validated. They are primarily leading indicators that warn of potential degradation in business outcomes before that degradation becomes visible in business KPIs. A model whose prediction confidence distribution has shifted significantly from its deployment baseline may not yet be producing more errors. Still, it is in a condition that historically precedes increases in error rates. Model health metrics allow proactive intervention rather than reactive incident response.

The core model health metrics are prediction volume, confidence distribution, data freshness, and drift index. Prediction volume confirms that the model is being used at the expected rate; a sudden drop in volume may indicate a deployment issue or a change in how the model is being surfaced to users. The confidence distribution tracks how certain the model is about its predictions compared to deployment. A shift toward lower confidence across the distribution signals that the model is encountering inputs it was not trained to handle. Data freshness measures whether the features the model consumes are being updated at the expected frequency. The drift index measures how much the input distribution has changed relative to the training distribution.

Model health metrics should be visible to the product and platform teams daily and summarized for business unit leaders weekly or monthly as part of the operational review. Business unit leaders do not need to understand the technical interpretation of every metric. Still, they should understand the health status of each model, green, amber, or red, and what each status means for the model's reliability. A traffic-light health summary is more actionable for a business unit leader than a set of technical statistics with no contextual interpretation.

Learning Metrics

Learning metrics quantify the rate at which the AI system improves over time. They bridge the gap between model health, which describes the current state, and model trajectory, which indicates whether the system is improving, staying the same, or declining. For managers making investment decisions, trajectory is at least as important as the current state. A model with mediocre current performance that is improving rapidly may be a better investment than a model with strong current performance that has plateaued.

The primary learning metrics are annotation yield (discussed in detail in the HITL chapter), model improvement rate per retraining cohort, and the proportion of production errors captured by the HITL system and fed into the improvement cycle. This last metric is particularly important because it measures the completeness of the feedback loop. A model that captures 50% of its production errors in the HITL review process is learning from half of its mistakes. A model that captures ninety percent is learning from nearly all of them. The difference in improvement trajectory over twelve months is substantial.

Learning metrics also inform investment decisions in retraining. If a model has been in production for two years and its improvement rate per annotation cohort has flattened to near zero, additional investment in HITL volume is unlikely to yield meaningful improvements. The returns from learning investment are diminishing. The organization should either accept the model as-is, invest in a more fundamental model-improvement effort (new data, new model architecture, new features), or begin evaluating whether to sunset the capability and replace it with a different approach. Learning metrics makes these investment decisions evidence-based rather than intuitive.

Linking Metrics Through Attribution

Attribution is the mechanism that links model behavior to what the model predicted and how confident it was, to downstream business outcomes. Without attribution, the organization knows the model is technically performing but does not know whether that technical performance translates into business value. Attribution closes that loop.

The simplest attribution approach is a before-and-after comparison: measure the business KPI before and after deployment, controlling other factors that changed during the same period. This approach works well when the deployment was a clear boundary event, when the model was turned on for all users on a specific date, and when no other significant changes to the workflow occurred simultaneously. In practice, both conditions are often violated, which is why more sophisticated attribution approaches are sometimes necessary.

A difference-in-differences design compares the change in business KPIs for users with access to the AI feature against the change for users without access, controlling for pre-deployment differences between the two groups. This approach requires a natural comparison group: a region, a team, or a product line that does not have access to the feature. This is not always available, but it can often be designed into the deployment rollout if the measurement architecture is planned. Staged rollouts, in which the feature is deployed to one group before another, naturally create a comparison group and should be the default deployment strategy whenever feasible.

The attribution methodology should be documented and reviewed by finance and business-unit leadership before it is used to make portfolio decisions. If the organization is going to sunset a capability because its attributed value is below the threshold, the business unit that owns it should understand and accept the attribution methodology. Attribution disputes after the fact, "you're measuring the wrong thing, " or 'that metric doesn't capture the full value, " are legitimate governance conversations that should happen before the portfolio review, not as a last-ditch defense during it. Investing in upfront alignment of attribution methodologies is cheaper than managing attribution disputes during a portfolio reallocation.

9.4 The Executive View

The minimal impact dashboard described in the previous section is designed for the product and operations teams who manage AI-assisted features day-to-day. Executives, the C-suite, the board, and the audit committee need a different view. They need a summary that answers their questions: Is the AI program delivering value? Are the risks under control? Are we getting better over time? Is there anything we need to decide?

Building an executive view requires translation, not simplification. Technical metrics must be expressed in business language. Risk indicators must be contextualized against risk appetite. Progress must be expressed against committed targets, not against what the team wishes the targets were. Executives are not served by optimism; they are served by accuracy that allows them to make good decisions. The executive view should be designed with that standard in mind.

One-Page Summary

The one-page executive summary is the single most important communication artifact in the AI measurement architecture. It must answer the executive's primary questions in less than 3 minutes of reading time, support a 10-minute conversation, and provide sufficient context for a decision on any item that requires executive input. Every word and every number on the page should be there because it serves one of those purposes.

A well-designed one-page summary contains four sections. The portfolio health section provides a traffic-light summary of each AI-assisted capability in production, green for on target, amber for at risk, and red for requiring immediate attention, with one sentence of context for each amber or red item. The financial section shows AI program costs versus budget, attributed business value for the period, and net value contribution. The risk section highlights any governance, compliance, or operational risk items that require executive awareness. The decisions section lists all items that require executive input or approval, with sufficient context to enable a decision.

The one-page summary should be produced on a regular cadence; monthly is appropriate for most organizations. It should be consistent in format across cadences so that executives can quickly identify changes from the previous period. Inconsistent formatting forces the reader to reorient with each report, consuming cognitive capacity that should be spent on the content. Consistency also makes it easier to detect trends across multiple periods, which is where the most important management signals often appear.

The discipline of producing a one-page summary every month forces the management team to maintain a clear, current understanding of the portfolio. Teams that struggle to produce a summary, lack current attributed value data, or cannot clearly articulate the status of every capability are uncovering measurement gaps that need to be addressed. The summary is both a communication tool and an instrument of management discipline.

Diagram 9.2 – Executive One-Page Summary Template

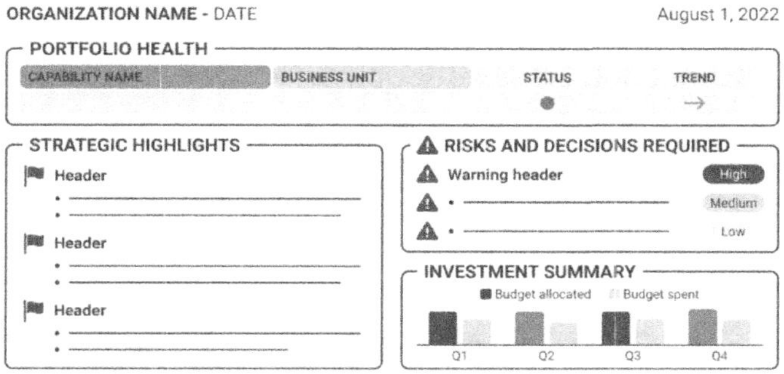

A structured one-page executive summary communicates AI portfolio health, highlights, risks, and investment.

Drilldown Architecture

The one-page summary creates the questions; the drilldown architecture answers them. A drilldown is a next-level view of any item on the summary that an executive or senior manager wants to understand in more depth. It should be available on demand without requiring a separate report request and organized to answer the specific questions the summary is likely to raise.

An effective drilldown architecture has three levels. The first level is the portfolio drilldown: a view of all active AI capabilities, including their KPI trends, health status, and financial performance over a rolling 12-month period. This level supports the portfolio management conversations described in the next section. The second level is the capability drilldown: a detailed view of a single AI feature that shows its business KPI trends, model health metrics, HITL metrics, attribution analysis, and operational costs. This level supports investment and risk decisions about a specific capability. The third level is the incident drilldown: a view of any active or recent incidents with timeline, scope, response actions, and resolution status. This level supports conversations about governance and accountability.

Drilldown architecture should be built on the same data that powers the minimal-impact dashboard, using the same metrics, attribution methodology, and time periods. Inconsistency between dashboard data and drilldown data is a credibility problem. When a business unit leader asks, 'How does this reconcile with what I saw on the dashboard last week?' and the answer requires a complex explanation that draws on multiple data sources and methodologies, the management team loses confidence in both views. A single, consistent data layer is not just a technical convenience; it is a governance requirement.

9.5 Portfolio Management

Portfolio management is the organizational practice of treating the collection of AI-assisted capabilities as a managed investment portfolio rather than as a set of independent projects. It applies the logic of investment management return optimization, risk diversification, rebalancing, and exit discipline to AI program strategy. Organizations that manage their AI capabilities as a portfolio consistently extract more value from the same level of investment than organizations that manage each capability independently.

The core portfolio management practices are the quarterly portfolio review, the sunset decision matrix, and the reinvestment process. Together, they create a governance cadence that ensures investment is continuously directed toward the highest-value opportunities, that underperforming capabilities are exited before they consume resources indefinitely, and that the lessons from previous investments are systematically applied to new ones.

Diagram 9.3 – Portfolio Review Decision Flow

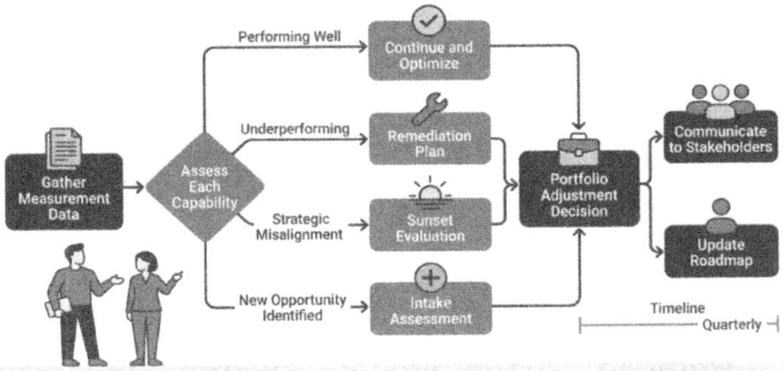

Quarterly portfolio reviews assess each AI capability and drive four possible strategic outcomes.

Quarterly Portfolio Reviews

The quarterly portfolio review is a structured, time-boxed session in which senior business and technology leadership reviews the performance of every AI-assisted capability in production and makes explicit investment decisions. It is not an operational status review; it is a strategic investment review. The questions it answers are not 'what happened this quarter?' but 'given what happened this quarter, where should we invest next quarter?'

The review agenda follows a standard structure. First, a brief presentation of the portfolio health dashboard traffic-light status, financial summary, and any active risk items to establish shared situational awareness. Second, a deeper review of any amber or red capabilities, with a decision about the path forward for each: improve, accept, or sunset. Third, a review of the investment pipeline use cases in development or in the backlog, and a prioritization decision on which to advance and which to defer. Fourth, a brief session on lessons from the quarter: what the portfolio review revealed that the organization did not know before, and what changes to measurement, governance, or investment process those lessons suggest.

The portfolio review should have explicit decision rights. The chair, typically the CIO, CAIO, or equivalent, has the authority to approve investment changes, initiate sunset decisions, and escalate items requiring broader executive input. Business unit leaders have the authority to accept or reject investment proposals that affect their areas. Engineering and data science leaders have the authority to commit or decline resource allocations within their teams' capacity. When decision rights are unclear, portfolio reviews become discussion forums without outputs, precisely the organizational pathology the insurer in this chapter's scenario suffered for three years.

A portfolio review produces three deliverables: an updated portfolio health summary, a set of investment decisions with owners and timelines, and an action log that tracks commitments made in previous reviews and their completion status. The action log is the accountability mechanism that prevents portfolio reviews from being quarterly discussions without consequences. If a team committed to improving the attribution methodology for a specific capability at the last review but has not done so, that should be visible and discussed in the current review. Accountability in portfolio management is structural, not personal; it is built into the process rather than dependent on individual diligence.

The Sunset Decision Matrix

A sunset decision matrix is a structured tool for determining whether to continue, modify, or retire an AI capability. It evaluates each capability against two dimensions: measured business value (the attributed impact on business KPIs relative to cost) and strategic alignment (the degree to which the capability supports the organization's current strategic priorities). The intersection of these two dimensions determines the recommended action.

High business value and high strategic alignment capabilities are the portfolio's core assets. They should be maintained and, where feasible, expanded. High business value but low strategic alignment capabilities, a capability that delivers measurable returns but in an area the organization is deemphasizing, should be maintained at current investment levels while a transition plan is developed. Low business value but high strategic alignment capabilities are candidates for redesign: the strategic direction is correct, but the current implementation is not delivering on it. Low business value and low strategic alignment capabilities should be sunset.

The sunset decision matrix is valuable because it makes the sunset decision a logical outcome of a transparent evaluation rather than a judgment call under political pressure. Business units that have invested in AI capabilities will naturally resist sunsetting them, even when the capabilities are not delivering value. A matrix-based process that evaluates capabilities against objective, agreed-upon criteria makes the conversation about whether the criteria are being applied correctly, rather than about whether the sunset decision is fair. The former conversation is much more productive than the latter.

Sunsetting a capability requires a migration plan and a customer communication strategy. If internal users have integrated the capability into their workflows, they need time and support to adjust. If external users, customers, partners, or regulators have been told the capability exists, they need to be informed about its retirement and given sufficient notice to adapt. Sunsetting without adequate communication is a trust and compliance risk. The migration plan and communication strategy should be developed as part of the sunset decision, not as an afterthought after the decision has been announced.

Diagram 9.4 – Sunset Decision Matrix

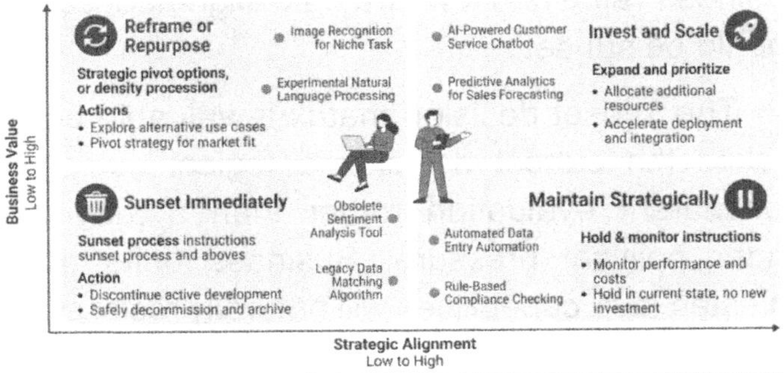

Use the sunset matrix to objectively decide whether to scale, repurpose, maintain, or retire AI capabilities.

Reinvestment and Capacity Reallocation

Sunsetting capability is only half the portfolio management decision. The other half is reinvestment: what does the organization do with the resources, engineering capacity, infrastructure spend, and business stakeholder attention that the sunset frees up? Organizations that sunset capabilities without a clear reinvestment plan often find that the freed resources are absorbed by maintenance work elsewhere rather than directed to higher-value opportunities.

Reinvestment decisions should be made at the same portfolio review session as sunset decisions, not in a separate process. If the insurer in this chapter's opening scenario had waited to reallocate resources until after the sunsetting process was complete, which can take months as migration plans are executed. Capabilities are wound down; the strategic opportunity created by the portfolio review would have been partially dissipated. Making the reinvestment decision simultaneously with the sunset decision creates a clear line of accountability: the released resources are already committed to specific higher-value uses before the sunset begins.

Reinvestment decisions should be documented in the portfolio review action log with specific resource commitments, responsible owners, and expected value targets. The expected value targets become the business KPIs against which the new investment will be evaluated in future portfolio reviews, closing the loop from reinvestment decision to measurement to portfolio review. Without that closing loop, the organization is executing a cycle without feedback, which means it will make the same kinds of reinvestment decisions indefinitely, rather than learning which types of investments consistently outperform and directing more resources to them over time.

Capacity reallocation also applies to non-financial resources: leadership attention, data stewardship capacity, reviewer time in HITL systems, and business unit collaboration. These resources are often more constrained than a financial budget, and their reallocation has a more immediate impact on the likelihood of a new initiative's success. A reinvestment plan that identifies specific leadership sponsorship and business unit partnership commitments, not just budget, is more likely to deliver on its value targets than one that allocates budget without considering the organizational resources that make budget productive.

9.6 Institutionalizing Learning

Portfolio discipline creates operational clarity about where the organization is investing and what return it is getting. But durable improvement in AI program effectiveness requires something more: the systematic capture and application of organizational learning. Every AI initiative the organization undertakes, whether successful or not, generates knowledge about what works in this organization's context, with its data, in its regulatory environment. That knowledge is enormously valuable. Most organizations let most of it evaporate.

Institutionalizing learning means creating the structures, incentives, and practices that cause organizational knowledge to accumulate rather than dissipate. It means building playbook libraries that capture the patterns from successful deployments and the lessons from failures. It means measuring the organization's learning rate, not just the performance of individual AI capabilities, but the rate at which new capabilities benefit from the organization's accumulated experience. And it means embedding learning practices into the governance cadence so they happen regularly, rather than only when someone finds time.

Diagram 9.5 – Institutional Learning Architecture

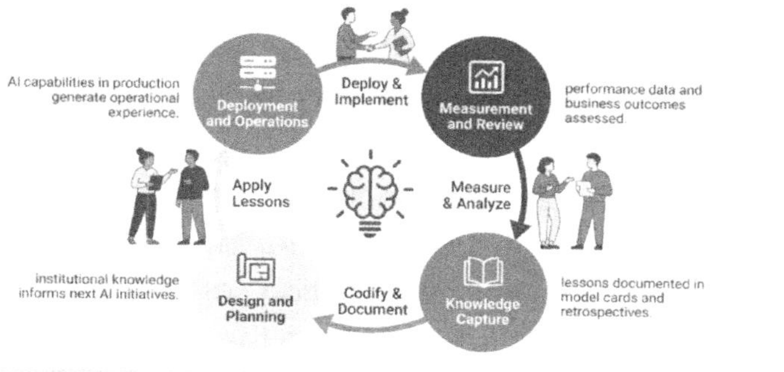

A continuous four-node learning cycle turns operational experience into organizational AI intelligence.

Playbook Libraries and Postmortems

A playbook library is a curated collection of documented patterns, decision frameworks, and lessons learned from the organization's AI deployment experience. It is not a documentation archive; it is a management knowledge base designed to make the next initiative faster, better, and less likely to repeat the mistakes of previous ones. The distinction matters because archives are organized for completeness; playbook libraries are organized for usefulness.

Effective playbooks are written at the level of decision, not at the level of implementation detail. A playbook for selecting a HITL pattern does not explain how to implement a confidence threshold in code; it explains which factors should influence the threshold-setting decision, what typical thresholds look like across different use cases, and what the organization has learned from its own implementations to watch for during calibration. An engineering team member who reads it should leave with better judgment, not just more information.

Postmortems are the primary input mechanism for the playbook library. A postmortem is a structured after-action review conducted after any significant event, such as a successful deployment, a production incident, a project cancellation, or a portfolio review finding, to extract lessons from the event. It documents them in a form accessible to others. Postmortems should be blameless in spirit; they are about understanding what happened and why, not about assigning faults, but they should be rigorous in practice. A postmortem that concludes with 'things went well' or 'we should communicate better next time' is not a postmortem; it is a summary meeting.

A specific function should own the playbook library, the CAIO's office, the platform team, or a dedicated AI program management function, with a defined contribution process and a regular review cycle. Playbooks that are never updated become outdated faster than they are consulted, which destroys confidence in the library as a resource. A semi-annual review cycle that flags each playbook as current, under review, or deprecated is a minimal quality-maintenance mechanism. Organizations that invest more in playbook quality maintenance by assigning specific subject-matter experts as playbook owners who review and update their playbooks after each relevant deployment will build a genuinely valuable management knowledge base over time.

Institutional Metrics

Institutional metrics measure the organization's AI governance capability, not the performance of individual AI systems, but the health and maturity of the organizational practices that govern them. They are the metrics that answer the question: Is this organization getting better at managing AI over time? They complement the capability-level metrics in the minimal impact dashboard, which measure whether individual AI features are working, with a view to the organization's own maturation.

The most valuable institutional metrics include: mean time to recovery for AI-related incidents (MTTR), which measures the organization's ability to respond to and resolve operational problems with AI systems; the percentage of active AI initiatives that have documented measurement architectures and attributed business KPIs, which measures governance discipline; the percentage of AI initiatives that have documented playbook contributions postmortems or pattern captures that have been added to the library; and the reuse rate for platform capabilities and governance frameworks, which measures whether the organization is actually compounding on its prior investments or rebuilding from scratch with each new initiative.

MTTR deserves particular attention as an institutional metric because it directly measures the organization's operational resilience. A low MTTR indicates that the organization has the runbooks, communication protocols, rollback procedures, and clear escalation paths to respond to incidents quickly and predictably. A high MTTR indicates that incident response is improvised, that the necessary tools and processes are not in place, or that accountability for incident response is unclear. MTTR trends over time are more informative than point-in-time values. An organization that systematically reduces its MTTR across all AI system categories is building operational resilience regardless of its starting point.

The percentage of initiatives with playbook contributions measures organizational generosity, the degree to which teams invest in making their experience available to others rather than hoarding it. This metric is worth tracking explicitly because the incentives do not naturally favor knowledge contribution. Teams are busy, and documenting what they learned takes time that could be spent on the next initiative. Leadership's attention to these metrics signals that knowledge contribution is valued and expected, not optional. Organizations that consistently achieve high playbook contribution rates build institutional knowledge at a rate that provides a meaningful competitive advantage over time.

Continuous Governance Cadence

Governance cadence is the rhythm of structured management attention that keeps the organization's AI program on track between strategic decisions. It encompasses daily operational reviews, weekly status cadences, monthly portfolio health reviews, quarterly portfolio investment reviews, and annual strategic planning cycles. Each cadence serves a different management purpose, and the ensemble of cadences creates a governance structure that is both responsive to immediate issues and oriented toward long-term performance.

The product and platform teams own the daily operational cadence. It covers model health dashboards, queue depth for HITL systems, and any active incidents. It is operationally focused and does not require business unit or executive participation, except during incidents. The weekly cadence is owned by product and operations leadership and covers HITL metrics, KPI trends, and any items escalated from the daily operational review. The CIO or CAIO owns the monthly cadence and covers the one-page executive summary, portfolio health, and any governance decisions that do not require a full quarterly review.

The quarterly review is the strategic cadence described in the portfolio management section: the full investment review, sunset decisions, and reinvestment allocation. The annual cycle is the strategic planning exercise that evaluates the AI program's direction against the organization's evolving strategic priorities, assesses the platform's capability maturity, and sets the investment thesis for the coming year. The annual cycle should produce a documented AI strategy that is visible to both internal stakeholders and, in appropriate form, to external stakeholders, including regulators and major customers.

Governance cadence only works if it is genuinely maintained. A quarterly portfolio review that is canceled because the calendar is full, a monthly one-page summary that is not produced because the data is not ready, or a postmortem that is never conducted because the team has moved on to the next initiative; each of these failures weakens the governance structure in ways that compound over time. Building governance cadence into standing calendar commitments, assigning clear ownership to each cadence event, and tracking cadence completion as an institutional metric are practices that distinguish organizations with durable governance from those with governance in name only.

The governance cadence is also the mechanism by which the organization adapts its AI practices to a changing environment. New regulatory guidance, new model capabilities, new risks surfaced by industry incidents, and new competitive pressures should all enter the governance cadence through a structured intake process and be assessed against current practices at the appropriate cadence level. An organization whose governance cadence is intact can absorb these changes systematically. An organization whose governance cadence has lapsed must respond reactively, which is more expensive and error-prone.

9.7 Manager's Checklist

Use this checklist to assess whether your AI program's measurement architecture and learning practices are adequate to support durable impact.

- Business KPIs defined for every AI-assisted capability in production, expressed as specific, time-bounded, directional targets.
- Attribution methodology is documented and accepted by business unit leadership before portfolio review decisions are made based on it.
- Minimal impact dashboard operational with all four metric categories: business KPIs, model health, learning metrics, and attribution.
- One-page executive summary produced monthly and reviewed in a standing leadership forum.
- Drilldown architecture is available on demand for all three levels: portfolio, capability, and incident.
- Quarterly portfolio review on the calendar as a standing commitment with defined attendees, agenda, and decision rights.
- Sunset decision matrix developed and accepted by business and technology leadership as the evaluation framework for portfolio decisions.

- Reinvestment decisions made at the same session as sunset decisions, with resource commitments and value targets documented.
- Playbook library established with a defined contribution process and ownership model.
- Postmortem process defined and applied to all significant deployments, incidents, and portfolio decisions.
- MTTR tracked across all AI system categories and reviewed as a standing item in the quarterly portfolio review.
- Percentage of initiatives with documented measurement architectures tracked as a governance compliance metric.
- Percentage of initiatives with playbook contributions tracked and discussed at the leadership level.
- Governance cadence documented with ownership, standing calendar commitments, and completion tracking.
- Annual AI strategy produced, reviewed by executive leadership, and communicated to appropriate internal and external stakeholders.

9.8 Key Takeaway

Impact measurement is not the last step in an AI deployment; it is the foundation for every subsequent decision. The insurer whose story opens this chapter did not fail to build good AI capabilities; she failed to build the measurement architecture that would have allowed her to know which capabilities were good and to invest accordingly. When she built that architecture, she did not just recover lost value; she unlocked a portfolio management discipline that made future investments systematically better.

The framework in this chapter- the minimal impact dashboard, the executive view, portfolio reviews, the sunset matrix, and the institutional learning practices- constitutes the operating system for a mature AI program. It is what allows an organization to treat AI not as a series of individual projects with uncertain outcomes, but as a managed capability that compounds in value with each cycle of investment, measurement, learning, and reallocation.

The manager's role in this system is to maintain the discipline it requires: to insist on KPIs before deployment, to protect the governance cadence when calendars fill up, to act on portfolio review findings even when the findings are uncomfortable, and to treat knowledge contribution to the playbook library as a genuine organizational obligation. Mission-aligned AI governance is not a set of policies posted on an intranet; it is a set of practices maintained by people who understand what they are for and are committed to keeping them alive. This chapter, and this book, have been about equipping those people to do that work.

10 Conclusion

This book has advanced a simple yet demanding idea: AI guardrails are not optional safeguards for a few exotic systems; they are the operating system for any organization that wants to use AI at scale and remain trusted. As an IT manager, you sit in the uncomfortable but powerful space between ambition and accountability. You are close enough to technology to know what it can do, and close enough to the business to feel the risk when things go wrong. The chapters you have just worked through are a playbook for turning that tension into a durable advantage.

We began by discussing why guardrails are important today and how unmanaged AI can lead to technical, legal, and reputational risks. Next, we outlined governance structures that clarify ownership, followed by practical policies and technical safeguards that integrate into workflows. Human oversight and risk triage ensure focus where it's needed most. Finally, we emphasized monitoring, addressing drift, and incident response, preparing for inevitable issues by efficiently detecting, containing, and learning from them.

The second half of the book focused on building an AI program that can withstand regulatory, organizational, and engineering pressures. You saw how to design compliance once and localize it across jurisdictions, instead of chasing every new rule with a bespoke fix. You embedded guardrails in CI/CD pipelines, model registries, and data workflows so that compliance is enforced by infrastructure, not heroics. Finally, you stepped back to look at culture and maturity: how organizations move from ad hoc controls to an embedded, quietly confident posture where responsible AI is "how we work here."

As you close this book, the most important next step is not to "do everything" but to pick one or two concrete moves and execute them well. That might be creating a simple AI use-case register and risk triage, standing up a lightweight governance working group, or adding fairness and data-quality checks to a single pipeline. Once that works, expand. Guardrails, like any infrastructure, are built layer by layer: policy, process, tooling, culture. If you keep trying each investment back to mission, value, and risk, you'll find it easier to secure sponsorship and sustain momentum over the long term.

AI will only become more capable, more connected to your core operations, and more visible to regulators and the public. Organizations that treat guardrails as friction will spend the next decade reacting to patching incidents, scrambling for audits, and explaining failures. Organizations that treat guardrails as a strategic asset will move faster with more confidence because they know their systems are monitored, explainable, and defensible. Your role is to help your organization join the second group. You now have the frameworks, language, and patterns to do that. The rest is leadership: taking the first step, making the hard calls when trade-offs arise, and building a culture where responsible AI is not a slogan but a habit.

11 Referenced Content

Accenture. *AI Operating Models: Designing for Scale, Safety, and Speed.* Accenture Research, 2024. (CH1)

Agarwal, Rahul, and Priya Desai. "From Pilots to Platforms: Operating Models for Enterprise AI." *Journal of Organization Design*, vol. 13, no. 2, 2025, pp. 45–68. (CH1)

Allen, Taryn, et al. "Human-in-the-Loop at Scale: Governance Patterns for High-Stakes AI." *AI and Ethics*, vol. 5, no. 1, 2025, pp. 33–57. (CH8)

Amazon Web Services. *Building AI-Enabled Organizations on the Cloud: Operating Models and Guardrails.* Amazon Web Services, 2023. (CH1)

Baker, Jonathan, and Elise Romero. "Psychological Safety and AI Adoption: How Culture Shapes Experimentation." *Human Resource Development Quarterly*, vol. 35, no. 1, 2024, pp. 21–44. (CH3)

Bapna, Ravi, et al. "AI-Augmented Work: Evidence from Field Experiments in Knowledge Work." *Management Science*, vol. 70, no. 11, 2024, pp. 6123–6147. (CH4)

Bersin, Josh. *Workforce Transformation in the Age of Generative AI.* The Josh Bersin Company, 2024. (CH2)

Boston Consulting Group. *The AI-Powered Organization: Operating Models for End-to-End Transformation.* Boston Consulting Group, 2023. (CH1)

Bughin, Jacques, and Eric Hazan. *Rewiring Work: How Generative AI Changes Operating Models and Talent.* McKinsey Global Institute, 2023. (CH1)

Capgemini Research Institute. *The AI-First Enterprise: Scaling Responsible AI Across Functions.* Capgemini, 2024. (CH5)

Carroll, Erin, and David Green. "Learning Systems, Not Training Events: Continuous Capability Building for AI." *Development and Learning in Organizations*, vol. 38, no. 3, 2024, pp. 12–18. (CH4)

Cascio, Wayne F., and John W. Boudreau. "AI, Work, and the Future of Talent Management." *Human Resource Management*, vol. 63, no. 1, 2024, pp. 3–18. (CH6)

Center for Security and Emerging Technology. *AI and the Enterprise: Adoption, Risk, and Organizational Capacity*. Georgetown University, 2024. (CH9)

CIO Council. *Playbook for the Responsible Use of Generative AI in the Federal Government*. Executive Office of the President, 2024. (CH5)

Daugherty, Paul R., and H. James Wilson. *Working with AI: Real Stories of Human–Machine Collaboration*. Harvard Business Review Press, 2023. (CH3)

Deloitte. *State of AI in the Enterprise, 6th Edition: From Ambition to Accountability*. Deloitte Insights, 2024. (CH9)

European Commission. *Artificial Intelligence Act: Provisional Agreement Text*. Publications Office of the European Union, 2024. (CH8)

EY. *Human at the Center: Designing AI-Enabled Work and Operating Models*. EYQ, 2023. (CH2)

Fountaine, Tim, et al. "Building the AI-Powered Organization." *MIT Sloan Management Review*, vol. 65, no. 2, 2024, pp. 42–51. (CH1)

Gartner. *Top Strategic Technology Trends 2025: Agentic AI and the Future Operating Model*. Gartner Research, 2025. (CH1)

Gawer, Annabelle, and Michael Jacobides. "Platforms, Ecosystems, and AI: Rethinking Digital Infrastructure Strategy." *Long Range Planning*, vol. 57, no. 1, 2024, article 102347. (CH7)

Google Cloud. *Operating Models for Generative AI: Patterns for Platforms, Products, and Governance*. Google Cloud, 2024. (CH7)

Harvard Business Review Analytic Services. *Leading with AI: How Managers Turn Tools into Transformation*. Harvard Business Publishing, 2023. (CH5)

IBM Institute for Business Value. *AI Governance in Practice: From Principles to Operating Models*. IBM, 2024. (CH8)

IEEE. *IEEE 7000-2024: Standard for Addressing Ethical Concerns during System Design*. IEEE Standards Association, 2024. (CH8)

Kellogg, Katherine C., et al. "Algorithmic Management and the Future of Leadership." *Academy of Management Annals*, vol. 18, no. 1, 2024, pp. 1–39. (CH5)

KPMG. *Trust in Artificial Intelligence: Building Confidence in Enterprise AI Systems*. KPMG International, 2023. (CH8)

Lee, Kai-Fu, and Chen Qiufan. *AI 2045: Leadership, Power, and the New Work Order*. Currency, 2025. (CH10)

McKinsey & Company. *The State of AI in 2024: Generative AI's Second Year*. McKinsey Global Survey, 2024. (CH9)

Microsoft. *Responsible AI Standard, v2: Enabling Trustworthy AI Systems*. Microsoft Corporation, 2023. (CH8)

MIT Sloan Management Review. *The Future of the Augmented Workforce: AI, Skills, and Organizational Design*. MIT Sloan Management Review, Special Report, 2024. (CH2)

Mollick, Ethan. *Co-Intelligence: Living and Working with AI*. Portfolio, 2024. (CH3)

National Institute of Standards and Technology. *Artificial Intelligence Risk Management Framework (AI RMF 1.0)*. NIST, 2023. (CH8)

OECD. *OECD Framework for the Classification of AI Systems*. OECD Publishing, 2023. (CH1)

OpenAI. *Model Spec for Frontier AI Systems: Safety, Alignment, and Deployment*. OpenAI, 2024. (CH8)

PwC. *AI Business Maturity 2025: From Pilots to P&L Impact*. PwC Research, 2025. (CH9)

Raisch, Sebastian, and Sebastian Krakowski. "Artificial Intelligence and Management: The Automation–Augmentation Paradox." *Journal of Management Studies*, vol. 61, no. 2, 2024, pp. 257–283. (CH3)

Rockmann, Kevin W., et al. "Hybrid Intelligence: Designing Workflows for Human–AI Collaboration." *Organization Science*, vol. 35, no. 3, 2024, pp. 611–632. (CH8)

Sacavém, António, et al. "AI-Driven Leadership: Decision-Making, Competencies, and Ethical Challenges—A Systematic Review." *Administrative Sciences*, vol. 16, no. 4, 2026, article 173. (CH5)

Salesforce. *AI-Ready Organizations: Culture, Skills, and Trust in the Era of Generative AI*. Salesforce Research, 2024. (CH3)

Saxena, Anurag, and Mary Uhl-Bien. "Relational Leadership in the Age of AI: Rethinking Followership, Agency, and Control." *The Leadership Quarterly*, vol. 35, no. 2, 2024, article 101746. (CH3)

Schrage, Michael. *Recommendation Engines for Management: Metrics, Models, and Behavioral Change*. MIT Press, 2023. (CH9)

Shneiderman, Ben. "Human-Centered AI: A Governance Agenda for Reliable, Safe, and Trustworthy Systems." *AI and Society*, vol. 39, no. 3, 2024, pp. 765–781. (CH8)

Stanford Institute for Human-Centered AI. *AI Index Report 2025: Tracking Global AI Development*. Stanford University, 2025. (CH9)

Tambe, Prasanna, et al. "AI, Management Practices, and Organizational Performance." *Management Science*, vol. 70, no. 1, 2024, pp. 1–23. (CH9)

UK Information Commissioner's Office. *Explaining Decisions Made with AI: Guidance for Organizations*. ICO, updated 2023. (CH8)

UNESCO. *Recommendation on the Ethics of Artificial Intelligence: Implementation Guidelines*. UNESCO, 2023. (CH8)

U.S. Department of Labor. *Artificial Intelligence and the Workforce: Policy Considerations for Employers and Workers*. Office of the Assistant Secretary for Policy, 2024. (CH2)

U.S. Government Accountability Office. *Artificial Intelligence: Key Practices to Help Ensure Accountability and Responsible Use in Federal Agencies*. GAO-24-106543, 2024. (CH5)

World Bank. *AI for Development: Building Capabilities, Safeguards, and Institutions*. World Bank, 2024. (CH10)

World Economic Forum. *Empowering AI Leadership: A Boardroom Guide to Risk, Strategy, and Governance*. World Economic Forum, 2023. (CH5)

World Economic Forum. *Jobs of Tomorrow: Large Language Models and the Future of Work*. World Economic Forum, 2024. (CH6)

World Economic Forum. *Governing AI in Operations: From Principles to Practice*. World Economic Forum, 2025. (CH7)

Zhang, Lihua, and Marco Silva. "Designing Human-in-the-Loop Review Pipelines for Generative AI in Regulated Domains." *AI Magazine*, vol. 45, no. 1, 2024, pp. 77–95. (CH8)

Made in the USA
Monee, IL
04 November 2026

67386758R00154